FinOps for Snowflake

A Guide to Cloud Financial Optimization

Y V Ravi Kumar
Velu Natarajan
Parag Bhardwaj

Apress®

FinOps for Snowflake: A Guide to Cloud Financial Optimization

Y V Ravi Kumar
Irving, TX, USA

Velu Natarajan
Buffalo Grove, IL, USA

Parag Bhardwaj
Irving, TX, USA

ISBN-13 (pbk): 979-8-8688-1735-9　　　　ISBN-13 (electronic): 979-8-8688-1736-6
https://doi.org/10.1007/979-8-8688-1736-6

Copyright © 2025 by Y V Ravi Kumar, Velu Natarajan, and Parag Bhardwaj

This work is subject to copyright. All rights are reserved by the Publisher, whether the whole or part of the material is concerned, specifically the rights of translation, reprinting, reuse of illustrations, recitation, broadcasting, reproduction on microfilms or in any other physical way, and transmission or information storage and retrieval, electronic adaptation, computer software, or by similar or dissimilar methodology now known or hereafter developed.

Trademarked names, logos, and images may appear in this book. Rather than use a trademark symbol with every occurrence of a trademarked name, logo, or image we use the names, logos, and images only in an editorial fashion and to the benefit of the trademark owner, with no intention of infringement of the trademark.

The use in this publication of trade names, trademarks, service marks, and similar terms, even if they are not identified as such, is not to be taken as an expression of opinion as to whether or not they are subject to proprietary rights.

While the advice and information in this book are believed to be true and accurate at the date of publication, neither the authors nor the editors nor the publisher can accept any legal responsibility for any errors or omissions that may be made. The publisher makes no warranty, express or implied, with respect to the material contained herein.

　　Managing Director, Apress Media LLC: Welmoed Spahr
　　Acquisitions Editor: Shaul Elson
　　Editorial Project Manager: Gryffin Winkler
　　Copy Editor: Kezia Endsley

Cover designed by eStudioCalamar

Cover Photo by Katherine Carlyon on Unsplash (unsplash.com)

Distributed to the book trade worldwide by Springer Science+Business Media New York, 1 New York Plaza, New York, NY 10004. Phone 1-800-SPRINGER, fax (201) 348-4505, e-mail orders-ny@springer-sbm.com, or visit www.springeronline.com. Apress Media, LLC is a Delaware LLC and the sole member (owner) is Springer Science + Business Media Finance Inc (SSBM Finance Inc). SSBM Finance Inc is a **Delaware** corporation.

For information on translations, please e-mail booktranslations@springernature.com; for reprint, paperback, or audio rights, please e-mail bookpermissions@springernature.com.

Apress titles may be purchased in bulk for academic, corporate, or promotional use. eBook versions and licenses are also available for most titles. For more information, reference our Print and eBook Bulk Sales web page at http://www.apress.com/bulk-sales.

Any source code or other supplementary material referenced by the author in this book is available to readers on GitHub. For more detailed information, please visit https://www.apress.com/gp/services/source-code.

If disposing of this product, please recycle the paper

This book is dedicated to the Snowflake community and to all FinOps professionals focused on maximizing the value of their organization's Snowflake investment.

Table of Contents

About the Authors xix

About the Technical Reviewers xxiii

Acknowledgments xxv

Chapter 1: FinOps for Snowflake 1

    Overview 1

    Balancing Cost vs. Business Value 2

    Snowflake Architecture from 10,000 Feet 4

        Database Storage 5

        Query Processing (Compute) 5

        Cloud Services 6

    FinOps Introduction 7

        Benefits of Adopting FinOps 8

    Consumption Pricing Model 9

        Consumption Model Challenges 10

    Consumption Pricing Model: Pros and Cons 13

        Challenges Without FinOps 15

        How FinOps Addresses Those Challenges 15

        Advantages of FinOps Implementation 16

    FinOps for Snowflake 17

        Visibility and Transparency 18

        Control and Budgeting 18

        Optimize Resource Utilization 19

    A Comprehensive Approach to the FinOps Strategy 19

        Investing in Skills 21

        Understanding User Needs 22

TABLE OF CONTENTS

    KPIs (Key Performance Indicators) 22

    Culture of Spend Awareness 22

    Empowering Teams 23

    Tracking What Counts 23

    Continuous Improvement 23

    Gamification 23

Shared Accountability Model Challenges 24

The Cost Optimization Framework for Snowflake 26

    Incident Fire Fighting 27

    Cost Tracking using Query Tag 27

    Map Owners Accurately 27

    Get Everyone to Help 27

    Use Automation to Detect Outliers 28

    Use Automation to Create Self-Service Queries 28

    System Level Optimization 28

    Evolution of the Snowflake System 28

    Data Operations 29

    Fine-Tune AI-Based Suboptimal Patterns 29

Summary 29

Chapter 2: FinOps Adoption for Snowflake 31

Overview 31

FinOps Framework for Snowflake 32

Visibility and Transparency Phase 34

    Understanding the Visibility and Transparency Phase 35

    Cost Monitoring Tools and Methods 36

    Establishing Tagging and Governance Practices 37

    Promoting Cost Awareness Across Teams 38

    Establishing Metrics and KPIs for Visibility 39

    Best Practices for Establishing Visibility in FinOps for Snowflake 40

    Integrating FinOps into the Organizational Culture 41

Control and Budgeting Phase 42
Understanding the Control and Budgeting Phase 43
Establishing Budget Controls and Policies 44
Leveraging Automation for Cost Control 45
Monitoring and Enforcing Compliance 46
Enabling Cost Accountability and Chargeback/Showback 47
Continuous Improvement and Feedback Loops 48
Enhancing Collaboration and Communication 49

Optimization Phase 50
Query Profiling and Optimization 50
Implementing Automatic Clustering 51
Using Materialized Views for Efficiency 52
Enhancing Performance with Query Acceleration Service (QAS) 52
Leverage Time Travel and Data Retention Policies 53
Using Snowflake Query Caching 54

Summary 55

Chapter 3: Visibility and Transparency 57
Overview 57
Enhanced Reporting and Continuous Evolution 58
Empowering Application Owners 59
Senior Management and Wider Stakeholders 60

The Snowflake Pricing Model 60
Collection of Performance and Usage Data 61
Snowflake Service Cost 63
Compute Service Cost 64
Storage Service Cost 69
Data Transfer (Ingress or Egress) Service Cost 71
Priority Support Cost 72

Exploring Usage Costs 73
UI-Based Cost Management 73
Using Predefined Queries 77

TABLE OF CONTENTS

Compute Service Cost 77
Storage Service Cost 81
Data Transfer Cost 82
Priority Support Cost 83
Tagging and Cost Attribution 84
Query Tag... 84
Object Tags 86
Relative Cost Per Query 86
Exploring the Billing Usage Statement.. 88
How to Locate the Usage Statement 88
Reconcile Usage Statement 88
Summary... 89

Chapter 4: Taking Control of Your Snowflake Spend 91
Overview 91
Monitoring Usage and Setting Limits: The Foundations of Cost Control 92
Resource Monitors and Credit Quotas 93
Monitor Types and Granularity 93
Automated Actions and Intelligent Alerts.. 95
Best Practices for Sustainable Monitoring 96
Common Pitfalls to Avoid 97
The Role of Budgets in Cloud Cost Governance 97
Benefits of Budgeting and Alerts 98
How Budgeting Works in Snowflake.. 99
Account Budgets: Organization-Wide Oversight... 99
Custom Budgets: Granular Financial Management 100
Spending Limits and Alerts: Enforcing Financial Discipline.................................. 100
Best Practices for Budget Management 100
Advanced Controls and Future Considerations.. 101
Applying Chargeback and Showback Models 101
Understanding Chargeback and Showback Models 102
Implementing the Chargeback Model.. 102

Table of Contents

Implementing the Showback Model 103
Best Practices for Implementation 103
Driving Financial Accountability Through Visibility 104
Automating Warehouse Efficiency 104
Understanding Auto-Suspend and Auto-Resume 105
Key Benefits of Automation 105
Combine with Budgeting and Alerts 106
Balancing Performance and Cost 106
What Are Warehouse Sizes and Why Do They Matter? 106
Key Benefits of Right-Sizing Warehouses 107
How Warehouse Sizing Works 107
Warehouse Sizing: Making the Right Choice 108
Best Practices for Warehouse Sizing 108
The Role of Access Control in FinOps 109
Why Access Control Matters 110
How Access Control Works in Snowflake 110
Role-Based Access Control (RBAC): A Scalable Model 110
Granular Permissions: Precision in Access Control 111
The Principle of Least Privilege and Regular Permission Reviews 112
Monitoring and Auditing: Visibility into Access 112
Best Practices for Access Control in Snowflake 113
Summary 114

Chapter 5: Compute Optimization Strategies 115

Overview 115
MPP Compute: The Role of Parallelism 116
Warehouse Sizing Guidelines 118
Adjust Auto-suspend Settings 118
Adjust Multi-cluster Warehouse Settings 119
MAX_CONCURRENCY_LEVEL 123
STATEMENT_TIMEOUT_IN_SECONDS 124
Query Acceleration Service (QAS) for Added Compute 124

TABLE OF CONTENTS

    Use Data Scans for Sizing 128
  Warehouse Provisioned Capacity: Used vs. Wasted 130
    Consolidate Workload 132
  Warehouse Utilization Heatmap 134
  Query Attribute Cost 136
  Do Large Warehouses Cost More? 137
    How Query Composition Affects Warehouse Processing 138
    Impact on Query Performance 138
  Factors that Affect Warehouse Utilization 139
    Row Explosion: How to Prevent Query Performance Issues 139
    Disk Spilling (Paging) 143
    Out of Memory (OOM) 144
    High Scanners (Using Partitions Scanned as a Heuristic) 145
    Warehouse Caching 147
    Auto-Clustering 149
    UNION Without ALL 151
  Summary 152

Chapter 6: Workload Optimization Strategies 153
  Overview 153
  Identifying High-Cost Query Patterns 154
    Analyzing Query Patterns with Query Hash 154
    Analyzing Query Patterns with Query Parameterized Hash 155
    Why Query Patterns Matter 157
  Tracking Object Usage for Cost Management 158
    Why ACCESS_HISTORY Matters 158
    Find Table and View Usage 158
  Manage Operational Skew 160
    Scenario: Skewed Column 161
    System-Wide Impact 161
    Best Practices for Managing Skew 161
    Find Skewed Join and Aggregate Columns 162

How to Mitigate this Skew 163
Reducing Data Movement Using Snowpark 164
    Snowpark-Optimized Warehouses 165
    Snowpark Performance Best Practices 167
Choosing Between ETL and ELT 168
    ETL (Extract, Transform, Load) 168
    ELT (Extract, Load, Transform) 168
    Optimization Considerations 169
    Recommended Strategy 170
Break Down Complex Transformations 170
    Recommended Strategy 171
Designing the Consumption Layer 172
    Why the Consumption Layer Matters 173
    Key Design Principles 173
    Intra-Row Calculations 174
Summary 176

Chapter 7: Storage Optimization Strategies 177

Overview 177
Managing Database Storage Costs 178
Understanding Continuous Data Protection (CDP) 178
Components of Snowflake Storage Costs 180
Storage Optimization Strategies 181
    Time Travel and Fail-Safe 181
    Manage Short-lived Permanent Tables 184
    Managing Large, High-churn Tables 185
    Storage Impact of Reclustering 187
    Point-in-Time Cloning 189
    Manage Storage in Snowflake Stages 191
    Data/Table Cleanup Based on Usage 193
    Manage the Storage Lifecycle 197
    Workspace Governance and Cleanup 200

TABLE OF CONTENTS

The Data Lakehouse, a Distributed Architecture 201
Profiling Storage and Cost Use 203
    Table Usage Metrics 204
Storage Profiling Using UI 204
Storage Profiling Using Pre-Built Queries 205
Summary 207

Chapter 8: Data Modeling Optimization Strategies 209

Overview 209
Clustering Keys: A Foundation for Performance 210
    Estimate Cost of Enabling Automatic Clustering 211
    Track Updates on Clustered Tables 213
    Cluster Key Usage 214
    Determine Clustering Benefits 216
    Cost Metrics for Clustering 217
Materialized Views 217
    When to Create Materialized Views 218
    Usage Optimization 218
    Usage Metrics for Materialized Views 224
Search Optimization Service (SOS) 225
    Estimate Search Optimization Cost 225
    Usage Optimization 227
    Determine Search Optimization Benefits 230
    Usage Metrics for Search Optimization 231
Hybrid Tables for Mixed Workloads 232
    Hybrid Table Cost Components 233
    Usage Optimization 233
    Usage Metrics for Hybrid Tables 235
Establishing Soft Referential Integrity 237
    Skip Unnecessary Joins 237
    Eliminate Unneeded Data Scans 237
    Usage Optimization 238

Handling Semi-Structured Data 241
Advanced Metadata Engine 241
How JSON Extraction Impacts Queries 242
Tips for Optimizing Query Performance on JSON Data 243
Usage Optimization 244
Measure Performance Metrics 246
Semi-structured Data Dos and Don'ts 246
Denormalizing Fact/Dimension Tables 247
Why Denormalize Fact/Dim Tables? 247
Addressing Diminishing Returns 247
Summary 248

Chapter 9: Data Load Optimization Strategies 249
Overview 249
Data Load Optimization Strategies 250
Use COPY INTO with Optimal File Sizes 251
Leverage Snowpipe for Continuous Ingestion 251
Adopt External Tables for Large, Static Datasets 251
Use Apache Iceberg for Multi-Engine Access 252
Optimize File Formats 252
Partition and Cluster Strategically 252
Automate Metadata Refresh for External Tables 253
Monitor and Tune Load Performance 253
Avoid Redundant Transformations 253
Apply FinOps Principles 253
Batch Loading: Efficient Data Ingestion at Scale 254
Considerations 255
Continuous Data Ingestion with Snowpipe 255
Automating Data Loads with Snowpipe 256
SQL Example: Creating a Pipe for Snowpipe 258
Key Benefits of Using Snowpipe 258
Snowpipe Cost Model 259

TABLE OF CONTENTS

Dynamic Tables for Scalable Auto-Refreshing Pipelines 259
    What Is a Snowflake Dynamic Table? 260
    Key Features 260
    Why Use Dynamic Tables? 261
    Dynamic Table Example 261
    Optimization Tips for FinOps 262

Apache Iceberg Tables: Eliminating Data Redundancies 262
    The Problem: Redundant Data Movement 264
    The Solution: Iceberg as a Central Data Layer 264
    Building a Modern Iceberg-Based Data Pipeline 264
    Maximizing Efficiency with Iceberg 265
    Best Practices for Implementation 266

Tailoring Virtual Warehouse Size for Efficient Data Loading 266
    Why It Matters 267
    Methodology for Warehouse Sizing 268

Summary 276

Chapter 10: Cloud Services Optimization Strategies 277

Overview 277

Understanding the Cloud Service Layer Architecture 278

Monitoring Cloud Services Usage 280

Patterns Driving Cloud Services Usage 283
    Pattern 1: Copy Commands with Poor Selectivity 283
    Pattern 2: High Frequency DDL Operations or Cloning 284
    Pattern 3: High Frequency, Simple Queries 284
    Pattern 4: High Frequency Information_Schema Queries 284
    Pattern 5: Result Scan 285
    Pattern 6: High Frequency SHOW Commands 285
    Pattern 7: Single Row Inserts and Fragmented Schemas 285
    Pattern 8: Complex SQL Queries 286

Exploring Metadata Layers and Best Practices 286

Understanding Metadata Operations Throttling 288

Reducing Metadata Operations Throttling 291
Reducing Data Dictionary Size and Query Complexity 292
    Use ACCOUNT_USAGE 292
    Create a Copy of the Data Dictionary (DD) 293
    Use SHOW Commands 294
    Periodic Data Dictionary Cleanup 295
Summary 296

Chapter 11: Data Availability Optimization Strategies 297

Overview 297
    Snowflake Availability Service Level 299
Cost Components of Replication and Failover 299
    Total Cost 301
Egress Data Movement Optimization 301
Managing Replication Spend Effectively 303
    Monitor Snowflake Replication Activity 304
    Track Database-Level Replication Costs 305
    Audit Replication History Regularly 307
    Forecast Replication and Failover Expenses 308
    Align Replication Policies with Business Needs 309
    Optimize Object Selection for Replication 311
    Set Failover Testing Frequency Strategically 312
Storage Consumption of the Secondary Account 313
    What Drives Storage Costs on the Secondary Account? 313
Leverage Replication for Strategic Account Migration 315
Conceptual Example: Estimating Replication Costs 316
    Compute: Running the Replication Jobs 317
    Storage: Holding the Replicated Data 317
    Data Transfer: Moving Data Across Regions 317
    Total Monthly Replication Cost 317
Summary 318

TABLE OF CONTENTS

Chapter 12: Snowflake AI/ML Cost Management 319

Overview 319

Snowflake AI Features Credit Table... 320

Cortex AI Features and Cost Considerations 323

    AI Cost Considerations... 324

    AISQL Functions and Cost Considerations 326

Optimizing Token Usage in Cortex AI.. 327

Comparing Token Usage Across Cortex AI Models 330

    Token Usage Comparison 330

    Impact of Guardrails on Token Usage 331

Monitoring AI Services Credit Usage... 331

    Daily AI Services Credit Usage.. 331

    Detailed LLM Function Usage 332

    Credit and Token Consumption by Query... 332

    Daily Token Usage by Model 333

Snowflake ML Capabilities and Cost Consideration 334

Building Cost-Effective ML Models 334

    Why GPU Acceleration Matters 335

    Snowflake Notebooks and Container Services.. 335

    Best Practices for GPU Optimization.. 335

    Cost Considerations... 335

    Monitor Snowpark Container Service Usage... 336

Cost Insight: Forecast Snowflake Usage 338

    Understanding the Data... 338

    Step 1: Prepare the Data 339

    Step 2: Create the Forecast Model 339

    Step 3: Save and Analyze the Forecast 340

Cost Insight: LLM-Powered Queries 341

    Example Use Cases 342

Summary.. 344

Chapter 13: What's New in Snowflake Cost Optimization 347

Overview 347

Cost Efficiency Through Snowflake Cortex AI 348

    AI Features Supporting Cost Efficiency 350

Gen2 Standard Warehouses 351

    What Are Gen2 Standard Warehouses? 352

    Key Benefits for Cost Governance 352

    Gen2 Warehouse Credit Table 353

Adaptive Compute for Smarter Workload Management 354

    What Is Adaptive Compute? 355

    Why Adaptive Compute Matters for FinOps 356

    Real-World Impact: The Experience of Pfizer 358

    Considerations and Limitations 358

Spend Anomalies with Real-Time Notification 359

    What Are Cost Anomalies? 359

    Example Use Case 360

Cost Management: Cross Account Cost View 361

Egress Replication Cost Optimization 362

Summary 363

Index 365

About the Authors

Y. V. Ravi Kumar is an extraordinary ability (EB1-A) Einstein Visa recipient from the United States, an Oracle certified master, an Oracle ACE director alum, a Snowflake data superhero, a Snowflake squad member, a coauthor, a mentor, a technical blogger, a frequent speaker at international conferences, and a technical reviewer. He is TOGAF certified, has published more than 100 technical articles, and is an Oracle open/cloud speaker. He is an IEEE senior member with more than 27 years of multinational leadership experience in the United States, Seychelles, and India in the banking, financial services, and insurance (BFSI) verticals.

He has coauthored seven books: *Oracle Database Upgrade and Migration Methods, Oracle High Availability, Disaster Recovery, and Cloud Services, Oracle GoldenGate with MicroServices, Oracle Cloud Infrastructure (OCI) GoldenGate, Oracle Global Data Services for Mission-Critical Systems, Mastering MySQL Administration,* and *Mastering PostgreSQL Administration*. He has also served as technical reviewer on six books: *Oracle 19c AutoUpgrade Best Practices, Oracle Autonomous Database in Enterprise Architecture, End-to-End Observability with Grafana, Maximum Availability Architecture (MAA) with Oracle GoldenGate MicroServices in HUB Architecture, The Cloud Computing Journey: Design and Deploy Resilient and Secure Multi-cloud Systems with Practical Guidance,* and *Azure FinOps Essentials*.

Ravi is an Oracle Certified Professional (OCP) in Oracle 8i/9i/10g/11g/12c/19c/23ai, and he is also an Oracle Certified Expert (OCE) in Oracle GoldenGate, RAC, performance tuning, Oracle Cloud infrastructure, Terraform, and Oracle engineered systems (Exadata, ZDLRA, and ODA), as well as certified in Oracle security and maximum availability architecture (MAA). He is certified in Snowflake Pro, PostgreSQL, MySQL, and MySQL Heatwave platforms as well. He is also multi-cloud architect certified (OCI, AWS, and GCP).

ABOUT THE AUTHORS

He has published over 100 Oracle technology articles, including on Oracle Technology Network (OTN) in *ORAWORLD Magazine*, on UKOUG in *OTech Magazine*, and on Redgate. He has spoken four times at Oracle Cloud/OpenWorld (OOW) in Las Vegas and San Francisco. Over the last decade, he has delivered technical sessions at Oracle user groups: IOUG, OCW, NYOUG, AUSOUG, AIOUG, UTOUG, SOUG, DOUG, and Quest Oracle Community.

Ravi has designed data centers and implemented mission-critical databases for core banking solutions for two major central banks. Oracle Corporation has published his profile on their OCM list and in their "Spotlight on Success" stories.

Velu Natarajan is a seasoned data expert with over 20 years of experience architecting scalable, high-performance data and database solutions across a wide range of platforms. His passion lies in the relentless pursuit of modernizing data platforms, ensuring they are not only efficient and high-performing but also scalable and durable to meet the evolving needs of the industry. As an industry trailblazer, Velu has been an early advocate of Cloud FinOps frameworks, leveraging them to elevate business success and achieve notable cost savings.

As a leader in the database cloud center of excellence (CCoE), Velu plays a crucial role in driving his organization's cloud transformation journey. His deep knowledge of database practices enables him to lead initiatives that ensure seamless migrations, business continuity, and operational efficiency.

Velu's pioneering work with cloud FinOps frameworks has led to significant business success and cost savings. His application of the FinOps framework to control costs in cloud-based data platforms was recognized with the Discover President Award in 2024. Through effective collaboration with cross-functional teams to understand database usage, optimize performance, manage costs, and establish best practices, Velu developed self-service tools and processes.

A respected contributor to the data and financial operations community, Velu has shared his expertise at prestigious events such as Snowflake Summit 2024, where he discussed strategies for optimizing performance and costs on Snowflake, maximizing business value, reducing budget risk, and improving the user experience with the data tool. He also served as a technical reviewer for *Azure FinOps Essentials*, published by Apress.

Certified as an AWS certified database specialty, FinOps certified practitioner, and snowflake architect, Velu is renowned for his insights into data products and his commitment to excellence. Outside of work, he enjoys playing soccer and watching epic movies.

Parag Bhardwaj is a principal product manager and a recognized expert in FinOps, cloud governance, and enterprise cost optimization. With more than 18 years of experience in IT infrastructure and cloud architecture, he has played a key role in implementing large-scale FinOps and modernization initiatives for Fortune 500 companies, including advanced strategies for optimizing Snowflake costs.

He wrote *Azure FinOps Essentials* and has authored several research papers covering topics such as:

— Comparing cloud cost management options like Azure reserved instances and savings plans

— Innovative approaches to cloud governance using CIS benchmarks across multiple clouds—revamping IT infrastructure with Azure landing zones

— The importance of leadership in creating a cloud center of excellence (CCoE)

Parag often shares his insights at international FinOps conferences like FinOps X, where he addresses how to better connect engineering and finance through practical FinOps strategies.

About the Technical Reviewers

Krishnakumar Kunka Mohanram is a passionate and results-driven IT professional with over 20 years of experience in the information technology industry. He holds a bachelor's degree in computer science engineering and an MBA, combining strong technical expertise with business acumen. Throughout his career, he has taken on diverse roles across enterprise technology, data platforms, and cloud solutions.

Krishnakumar is a lifelong learner who constantly explores emerging technologies and applies them in real-world scenarios. He has delivered technical presentations at national and international conferences, and has actively contributed to various technical forums across India. As a coauthor of three technical books, his writing is known for being practical, insightful, and accessible to a wide range of audiences. His work has been well received by readers and recognized by global thought leaders.

In addition to authoring, he serves as a technical reviewer, ensuring content accuracy, clarity, and depth in specialized domains like data warehousing, FinOps, and cloud computing. He is also an IEEE senior member, reflecting his commitment to professional excellence and contributions to the engineering community.

Sanjay Kattimani has over 20 years of experience advising major financial institutions on scalable and robust enterprise architectures. He is widely recognized for his deep expertise in data platforms and cloud technologies teams and for contributing actively to the global tech community.

Dayakar Siramgari has over two decades of innovation in cloud architecture, data engineering, and GenAI platforms. Currently at T-Mobile, he leads transformative efforts in digital modernization, cost optimization, and enterprise-scale analytics. His cross-platform expertise spanning Azure, AWS, Snowflake, and Databricks enables him to architect future-ready systems across industries. With a unique blend of technical depth and visionary thinking, Dayakar's review ensures this book is both practical and cutting-edge.

Acknowledgments

I am grateful to God who gave us all the strength, courage, perseverance, and patience in this sincere and honest attempt of knowledge sharing. This eighth book of mine as a coauthor would not have been possible without the following people: Shri Yenugula Venkata Pathi and Smt. Yenugula Krishna Kumari, my parents, who instilled in me good thoughts and values, and Shri B. Suresh Kamath (founder of LaserSoft and Patterns Cognitive), my mentor, my guru, my strength, and my guide, who has inspired me for the last 27 years.

B. Suresh Kamath is an immensely talented and technically sound individual. He taught me how to be well read, with no compromises. He led by example in being content yet hungry for knowledge. He motivated me to go that extra mile in attempting and experimenting with newer technologies/environments and in being regularly out of my comfort zone.

Anitha Ravi Kumar, my wife, was immensely tolerant. "Behind every successful man, there is a good woman," as they say. I believe she is the embodiment of this well-known belief. Special thanks to my daughter, Sai Hansika, and my son, Sai Theeraz, for giving me time to write a eighth book in the past seven years.

I would like to thank Krishnakumar Mohanram Kunka, Dayakar Siramgari, and Sanjay Kattimani for accepting to be the technical reviewers for this book. Special thanks to Shaul Elson, Gryffin Winkler, Nirmal Selvaraj, Celestin Suresh John, and Laura Berendson at Apress for giving me the opportunity to write my eighth book for Apress. Thank you to the readers for picking up this book. We have attempted to be as simple and straightforward as possible when sharing this knowledge, and we truly believe that it will help you to steadily deep dive into various interesting concepts and procedures.

I would also like to thank the complete Infolob Global team—Vijay Cherukuri, Tim Fox, Josh Turnbull, Nivas Nadimpalli, and Satyendra Pasalapudi.

I would also like to thank the Snowflake Dallas User Group and Snowflake Community—Amilee (San Juan) Alesna, Joyce Avila, Anh Phuong Ta, and Aran Sakzenian—for their excellent support.

—**Venkata Ravi Kumar Yenugula**

ACKNOWLEDGMENTS

For many years, I aspired to share my experiences and insights with the community that shaped my journey and growth. Inspired by the words of the Greek philosopher Epictetus, "If you wish to be a writer, write." I finally embarked on the journey of writing my first book—an endeavor that has been both humbling and deeply fulfilling.

I am profoundly grateful to God for guiding me along this path. I bow in reverence to my parents, Shri. Natarajan Appadurai and Smt. Govindammal Natarajan, whose values and unwavering support have been the foundation of my life. I also extend heartfelt gratitude to my uncles and aunt—Kuppan E, Varadhan E, Mani E, Ramu E, and Valliammal M—whose sacrifices and encouragement have been a constant source of inspiration.

To my beloved wife, PadmaPriya Velu, the unwavering pillar of my life. Thank you for your unconditional love and support in everything I do. I am equally grateful to my children, Jyotsna Velu and Harshad Velu (Sai), for their understanding and for giving me the time and space to bring this book to life.

My sincere thanks to Nagesh Perumalla, Arvind Narayanamurthy, and Chandhar Ramalingam for their mentorship and valuable insights throughout this journey.

I am deeply grateful to the leadership and colleagues at Discover Financial Services (a Capital One division), including Vijay Gopal, Will Hinton, Ellen Klebanov, Vlad Meyman, Sarang Kale, and Dirk Anderson (Snowflake) for their instrumental role in shaping my professional journey in FinOps and for their support, which has been invaluable to this book.

In addition, I want to extend my heartfelt thanks to the leadership at GoodRx—Suresh Appavu and Andrew Douglas—for their trust and for providing me the opportunity to focus on FinOps and grow in this space.

I am honored to have collaborated with my coauthors—Y V Ravi Kumar and Parag Bhardwaj—and extend my appreciation to our technical reviewers, Krishnakumar Kunka Mohanram, Dayakar Siramgari, and Sanjay Kattimani, for their dedication and contributions to this project.

A special note of thanks to the team at Apress—Shaul Elson, Shobana Srinivasan, Gryffin Winkler, Nirmal Selvaraj, Celestin Suresh John, and Laura Berendson—for believing in us and giving us the opportunity to publish this book.

And finally, to you, the reader: thank you for picking up this book. We have strived to present our knowledge in a clear and practical manner, and we hope it helps you unlock greater business value from every dollar you spend on Snowflake.

—**Velu Natarajan**

ACKNOWLEDGMENTS

I am grateful to God who gave us all the strength, courage, perseverance, and patience in this sincere and honest attempt at knowledge sharing.

This book would not have been possible without the support and encouragement of the following people:

A heartfelt thank you to my parents, Radhey Shyam Sharma and Daya Sharma, who invested in me the values of hard work, integrity, and curiosity. Their blessings and wisdom continue to guide me through every endeavor, personally and professionally.

To my loving family—Anshu Bhardwaj, Nehal Bhardwaj, and Pranshu Bhardwaj—thank you for being my constant source of strength, joy, and inspiration. Your love and patience have enabled me to dedicate countless hours to this journey, often during late nights and weekends, and I am forever grateful for your unwavering belief in me.

My journey in the world of cloud has been shaped by close collaboration with FinOps and governance teams, where I've served as the product manager driving scalable solutions. I've had the opportunity to lead the development of governance frameworks and cost optimization capabilities that support enterprise-wide accountability. This book captures the real-world insights I've had when shaping products that empower financial transparency, spark meaningful strategic dialogue, and foster a culture of efficiency through data-informed decisions and cross-functional collaboration.

Special thanks to the technical reviewers—Krishnakumar Kunka Mohanram, Dayakar Siramgari, and Sanjay Kattimani—for their thoughtful feedback and insights throughout this journey. Their attention to detail and technical depth significantly strengthened the quality of this work.

To the Apress team—Shaul Elson, Gryffin Winkler, Nirmal Selvaraj, Celestin Suresh John, and Laura Berendson—thank you for allowing me to bring this work to life and for supporting our vision every step of the way.

Lastly, to the readers—thank you for picking up this book. We've done our best to present the content clearly, and we genuinely hope that it helps you navigate and deepen your understanding of the subject with confidence and curiosity. Whether you're just beginning your journey or are an experienced professional, we hope this book becomes a trusted companion during your learning journey.

—**Parag Bhardwaj**

CHAPTER 1

FinOps for Snowflake

Achieving the Perfect Balance Between Performance and Cost

Overview

Snowflake has transformed how organizations store, manage, and analyze data with its unique architecture and consumption-based pricing model. This technology scales horizontally and is the most flexible and fastest way of doing analytics with the most advanced cloud capabilities, including AI/ML. However, effective cost management in your cloud data platform requires a fresh perspective and systematic review to establish boundaries with platform consumers while advocating best practices for its effective use. Such duties demand detailed knowledge of the platform's usage and a thorough understanding of its architecture, along with proper dos and don'ts for a preventive stance in financial operations.

Financial operations, or *FinOps*, is the practice of wise spending and deriving maximum business value with shared responsibilities among multiple stakeholders in an organization. It is a framework of shared financial accountability with cloud operations, helping organizations manage their cloud spending efficiently and optimize their resources.

In this book, we explore essential strategies for effectively managing the cost in the Snowflake environment through the lens of the FinOps framework. This book provides best practices for managing spending well in Snowflake, focusing on reducing and optimizing cost redundancies, highlighting the importance and relevance of FinOps, and demonstrating how it helps organizations make wise financial decisions. Finally, we address common challenges in real-world situations and explain how FinOps principles can help tackle those challenges.

In the early days of cloud computing, there was no common terminology like "FinOps." It was previously referred to as "cloud cost management," "cloud cost optimization," or "cloud financial management." These terms were only later replaced with FinOps to reflect the cross-functional nature.

As we progress, you will discover how FinOps can transform the financial management of Snowflake and keep costs under control.

Welcome to the world of FinOps for Snowflake! Let us embark on this journey together.

Here is the gist of what we unleash about the essence of Snowflake FinOps in this chapter:

- Uncovering the essence of FinOps and understanding key terminology

- Reviewing Snowflake's architecture and getting acquainted with the core components and services that frame the platform

- Exploring the foundational principles of FinOps, including its framework and core components

- Discussing the consumption pricing model and shared accountability

- Highlighting the guidelines and best practices

- Showcasing the cost optimization framework and its impact on Snowflake cost optimization

Balancing Cost vs. Business Value

The cost of cloud-based systems is measured in fundamentally different ways from on-premises options. One critical difference is the "pay-as-you-go" model, where you only pay for the resources you use. This model provides flexibility and scalability, but it also introduces a new set of challenges. Stakeholders, such as the finance department, must become well-versed with the platforms and usage models to set boundaries and expectations with stakeholders.

Let's discuss some of these challenges with a few examples:

Understanding cloud costs: Organizations must train finance teams on the new operational model of cloud cost structures and the unique cloud invoicing processes. Due to dynamic pricing models and usage-based billing, predicting cloud costs can be challenging. Finance teams with a detailed understanding of the cloud cost portfolio can make more informed budgetary decisions in close collaboration with multiple stakeholders using the platform.

Finance team visibility (just-in-time cloud consumption cost): Finance teams need near real-time visibility into resource utilization to understand the financial implications of cloud service consumption. This knowledge allows them to compare current spending patterns with the overall economic strategy of the company and adjust either their spending or strategy accordingly.

Budgeting and forecasting: Traditional budgeting methods often fall short in an Agile cloud environment because cloud spending is inherently more dynamic and flexible. Unlike the fixed cost of on-premises infrastructures, which are almost always fixed, cloud expenses vary based on usage patterns (resource allocation and features used). This variability requires finance teams to adopt a more open-ended approach to planning and budgeting.

A well-structured FinOps model helps organizations navigate the complexities of cloud financial management. Companies that are new to Snowflake and want to focus on cloud spend, or companies with complex and large-scale Snowflake deployments should consider adopting a financial operations (FinOps) framework. This framework enables them to manage Snowflake costs collaboratively across various business functions, such as finance, technology, and operations.

As illustrated in Figure 1-1, FinOps brings a balance between cost and business value, helping organizations instill financial discipline in the cloud while maximizing the value they derive from it.

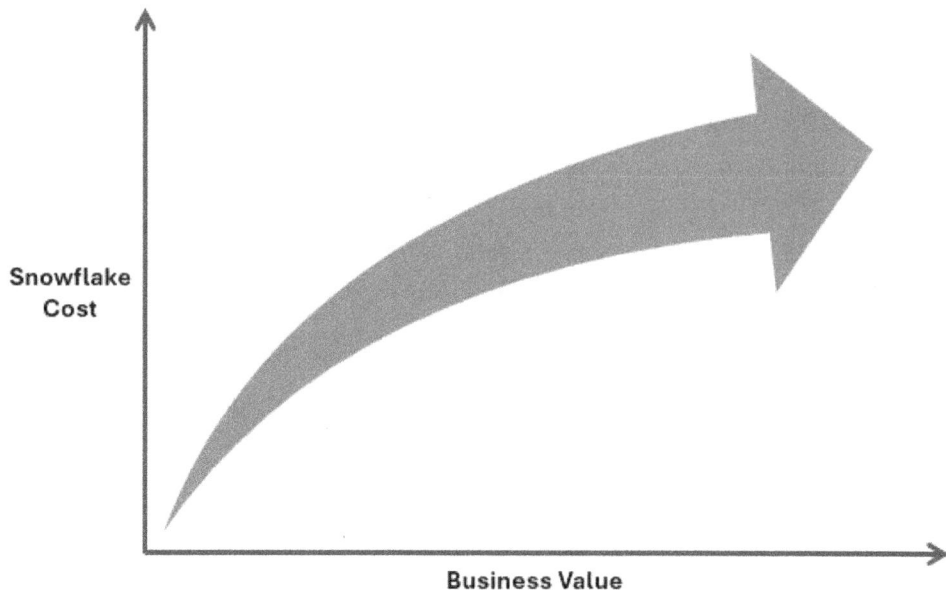

Figure 1-1. Balancing cost vs. business value

This book explores how FinOps addresses critical financial issues, introduces best practices, and becomes integral to modern financial operations for managing Snowflake costs.

Snowflake Architecture from 10,000 Feet

Snowflake uses a hybrid of traditional shared-disk architecture with a centralized data repository for persisted data and a shared-nothing architecture for query processing, enabling performance and scale-out capabilities. Built on a massively parallel processing (MPP) architecture, Snowflake can handle huge amounts of data simultaneously, delivering fast and efficient results. The shared-disk component stores persistent data in one centralized location, while the shared-nothing approach handles the heavy lifting of querying and processing. This hybrid design enables Snowflake to excel in large-scale data analytics. One of the key advantages of Snowflake's architecture is its ability to independently control storage and compute resources. This flexibility helps organizations manage costs effectively while maintaining high performance.

The next section examines at a high level how the Snowflake architecture processes each query and the factors that can accelerate or slow down these processes.

Database Storage

When you load data into Snowflake, it is transformed into Snowflake's internal optimized, compressed, columnar format and stored in cloud storage. Snowflake manages everything related to data organization, sizing, structuring, and compression, as well as metadata, statistics, and other aspects of data storage. The data objects maintained by Snowflake are not directly visible or accessible to customers; they are only accessible via SQL query operations executed using Snowflake.

Snowflake decouples storage and compute resources, allowing them to scale independently. This means you can increase computing power for complex queries without needing to increase storage capacity, and vice versa. This decoupling architecture also allows for simultaneous data reads and writes, eliminating the need for separate reporting and batch windows, unlike legacy architectures like Teradata, where separate windows for loading and reporting are required. This is a major transformation.

Query Processing (Compute)

Query processing in Snowflake is driven by virtual warehouses, which are independent compute clusters optimized for massively parallel processing (MPP). Each virtual warehouse has its own resources, ensuring consistent performance without resource contention. This structure also offers easy scaling: resources can be added or removed and scaled up or down, based on workload demands. This flexibility makes Snowflake efficient at handling various data processing needs, whether requiring more power for large tasks or less for smaller ones.

The architecture splits queries across nodes, enabling parallel processing of data. This allows for greater parallelism compared to systems that use a single thread or single dimension of parallelism. Efficiency is enhanced by intelligently distributing data throughout the nodes to avoid data movement and maximize speed and throughput. Each virtual warehouse is an isolated compute cluster that does not share resources with other virtual warehouses, ensuring no performance impact on other workloads.

With these capabilities, Snowflake's MPP architecture allows for concurrent execution of multiple workloads without performance impact. For example, an Extract, Transform, and Load (ETL) workload and an analytics workload can have their own isolated compute provisioned, even while operating on the same databases and tables. This means the ETL workload does not impact the performance of the analytics workload, and vice versa.

Cloud Services

Cloud services in Snowflake manage all activities across the platform. These services interlace various aspects of Snowflake to perform user requests, from login through query dispatch. Snowflake provisions compute instances from the cloud provider to run the cloud services layer.

The following services are managed in this layer:

- Authentication
- Infrastructure management
- Metadata management
- Parsing and optimization of query
- Access control

This flexible architecture of Snowflake, featuring elastic multi-cluster compute and optimized storage, as shown in Figure 1-2, allows it to efficiently handle large-scale data processing. This infrastructure can be seamlessly scaled up to meet growing demands. Additionally, the integration of AI (Artificial Intelligence) and ML (Machine Learning) capabilities enables models to run directly within Snowflake, minimizing the need for data movement and significantly enhancing performance.

CHAPTER 1　FINOPS FOR SNOWFLAKE

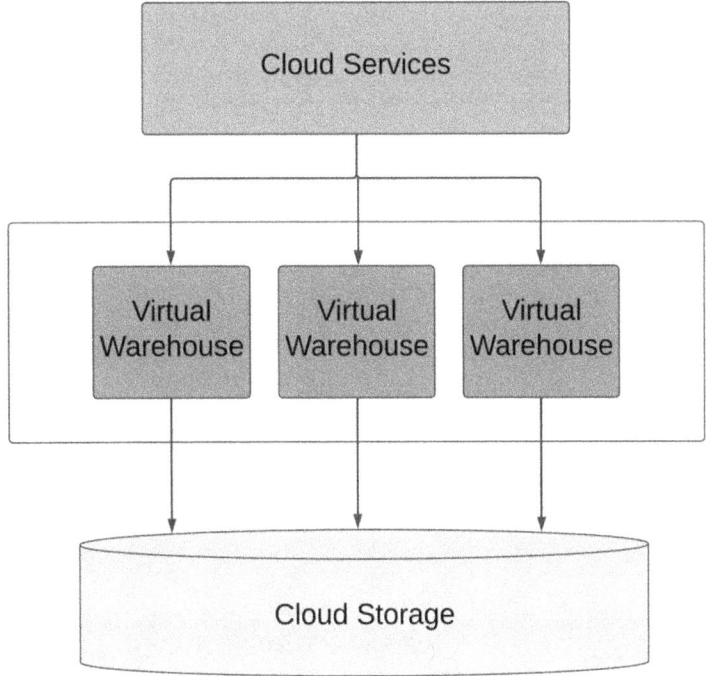

Figure 1-2. *Snowflake architecture from 10,000 feet*

Before discussing the concepts of FinOps, the next section explores how it evolved and the significant benefits it brings to modern cloud financial management

FinOps Introduction

In the world of AI and ML, data generation is growing faster than ever before. This data can help organizations identify new business opportunities, but they need effective analytics and management to make better investments. Since much of this data is unstructured, relational databases are not efficient in processing it. Cloud data platforms such as Snowflake, Databricks, Teradata, and AWS Redshift provide scalable storage and compute for high-velocity data generated by modern applications.

As more companies move to the cloud, solutions are needed to monitor and predict technology spending trends. FinOps is the emerging practice for rapid-cost-change environments, combining financial management and efficient cloud operations into one team. This results in earlier collaboration to contain costs. Due to increasingly fluctuating costs from multiple sources, many organizations are now adopting this trend of working with different disciplines.

7

In this context, having a strong FinOps strategy is more than just keeping track of cloud expenses. It's about balancing performance, cost, and quality. It is not the same as choosing the cheapest computing options to save money, as this can hurt the performance of essential tasks. Conversely, opting for high-quality resources can improve processing speed, but this often comes at a higher price. An effective FinOps methodology keeps organizations on track by using data to drive necessary decisions aligned with company goals and budgets.

FinOps is a framework with a set of capabilities, as shown in Figure 1-3. It enables teams within an organization to operate and collaborate to increase their understanding of cloud costs—understanding how much they have spent in the past, tracking today's expenditure, and predicting future spends. It helps them monitor these costs toward meaningful objectives and share what the tradeoffs are at every point along the way.

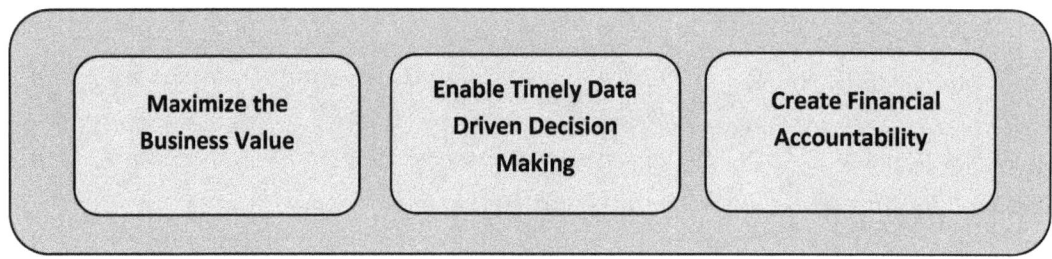

Figure 1-3. *FinOps capabilities*

FinOps is a framework for how teams within an organization can operate and collaborate to increase their understanding of cloud costs—how much you have spent in the past, what you are spending today, and what future spends are likely to be. It helps teams monitor these costs toward meaningful objectives and share what the tradeoffs are at every point along the way.

Benefits of Adopting FinOps

FinOps goes well beyond just cost reduction; it involves tuning resource utilization for performance excellence. With the FinOps framework in place, organizations can identify areas for improvement and move toward applying the best financial practices that everyone is aligned on, regardless of their team within the organization.

Collaboration: FinOps encourages open team communication, ensuring everyone speaks a common language regarding financial responsibilities. This atmosphere within teams enhances decision-making and builds accountable partnerships.

Proactive cost management: FinOps empowers organizations to proactively monitor potential financial risks—such as unexpected costs and budget overruns—through regular cost analysis and performance checks with near real-time data.

Continuous improvement: FinOps is a journey of continuous improvement. Companies periodically evaluate their spending habits and identify areas to optimize cloud resources and reduce costs.

Budget and limits: FinOps defines dynamic spending limits and budgets, making the impact of cloud investments on future resource allocations transparent.

Now that you have learned about the key benefits, the next section looks at the consumption pricing model. We take a closer look at how variable costs drive business decisions and explore the challenges that we need to handle.

Consumption Pricing Model

In a nutshell, you get the capacity you need to scale your business!

The pay-as-you-go model means that you are only charged for the cloud resources that you use. The benefit of the consumption-based model is that cloud resources can be provisioned and scaled on demand according to your requirements, without paying upfront and predictable instance billing first. The billing model—per resource, per second—on the cloud allows for fine-grained price control. It provides cost efficiency and elasticity in how businesses manage resources, with:

Variable costs: Unlike fixed costing models, consumption-based models can fluctuate based on actual use.

Resource explosion: This is when a business provisions more resources or services than required, resulting in unused or underutilized capacity. This doesn't happen with the pay-as-you-go model.

Scalability and flexibility: Users can scale their resource consumption up and down as dictated by business requirements, leading to higher control over usage and costs.

Snowflake has become the new norm for many enterprises ,as an increasing number of business organizations adopt consumption-based data infrastructure and services models. While the pay-as-you-go model offers flexibility and scaling capabilities, it also introduces complexities in the cost management process. FinOps enables granular visibility—a crucial discipline in today's cloud-first world. Granularity has become a cornerstone for cost management, financial accountability, and realizing value from Snowflake investments.

Consumption Model Challenges

Technology spending was not complex until organizations started migrating workloads to cloud-based solutions. As we all know, cloud computing is mandatory in modern software development because of its elasticity and scalability, which allows for fast-paced business adoption. It is crucial to ensure that cloud resources are not wasted by over-provisioning and under-utilizing. Figure 1-4 clearly shows the usage below the provisioned capacity, which incurs redundant and unnecessary costs. We explore the options to balance resource provisioning versus utilization, which is crucial to curtail redundant costs.

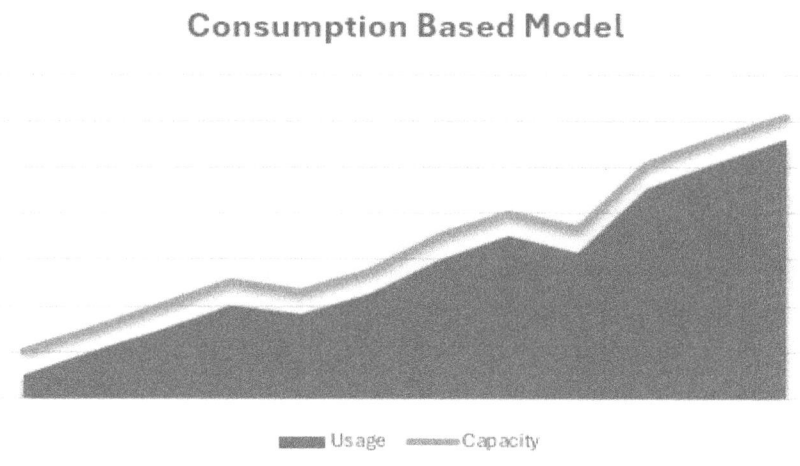

Figure 1-4. Consumption-based model

The benefit of effective usage comes with complications that must be navigated thoughtfully and strategically so that costs, resources, and the size of deliverables can be managed appropriately. The following sections discuss a few of the common challenges.

Variable Cost

In the cloud, costs vary between different resources and services. They also vary based on the region in which services are hosted. Such unpredictability makes budgeting and forecasting even more difficult.

New Operational Processes

Organizations have traditionally followed methods for budgeting that were useful for traditional solutions. However, these methods are inefficient for handling variable cloud expenses. In the cloud, budgeting should be flexible to adjust to changing use.

Increased Collaboration

With cloud costs growing increasingly complicated, the push for cross-functional engagement can drive value at an organizational level. Each team faces unique challenges and needs clearly defined practices for better control and clarity:

Finance teams: Under constant pressure to lower costs while supporting business goals, finance teams need to decide which expenses to approve and which to deny. They need tighter restrictions and bans on spending. Close collaboration with different consumer stakeholders is essential for justifiable budgeting.

Business units: All business units in the organization need to accelerate FinOps processes to improve customer service, responsiveness, and experience. Their decisions need to be agile in responding to market needs.

Engineering teams: Positioned in the middle, engineering teams balance cost efficiency with a focus on performance and innovation demands. They should work with finance and business units to ensure resources are utilized properly, costs are justified by business demands, and budgets are adhered to.

Figure 1-5 highlights the differences between generic cloud cost management and the FinOps cost management approach, along with the challenges in table format. As you have seen, FinOps stands apart from traditional cloud cost management by shifting from a single-team focus to cross-team collaboration. It also emphasizes usage forecasts and moves from reactive to proactive cost management strategies.

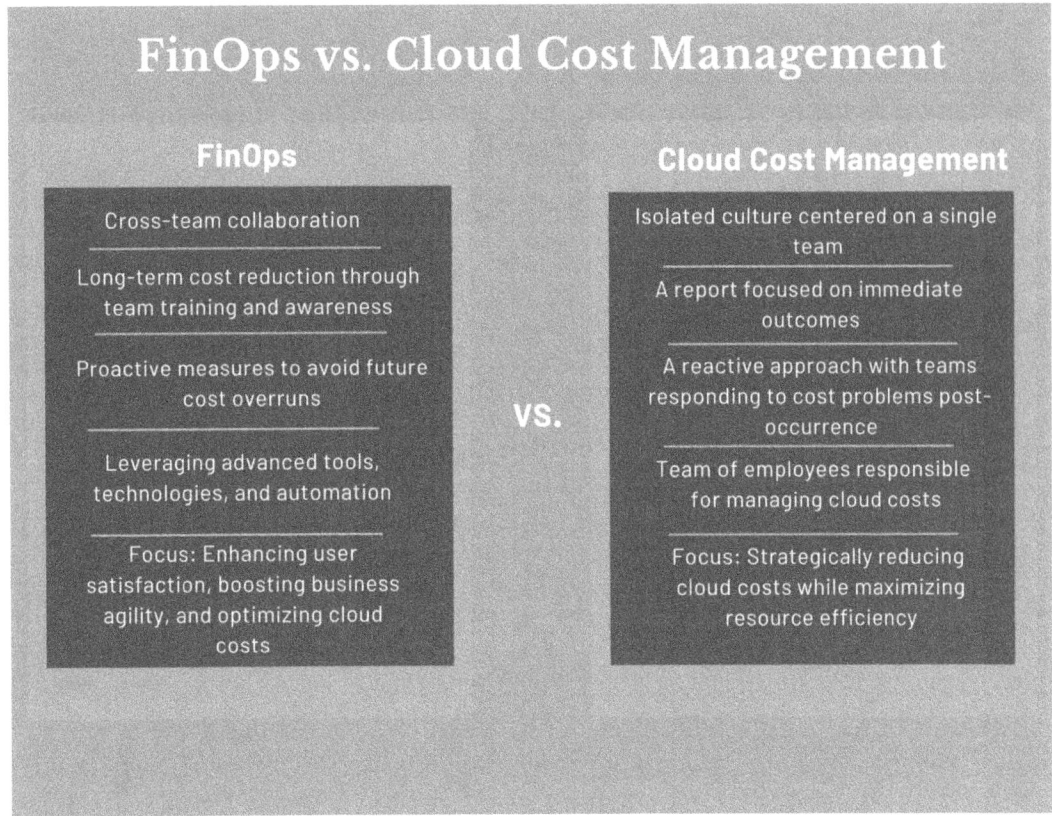

Figure 1-5. FinOps vs. cloud cost management

The next section covers the advantages and disadvantages of using the consumption pricing model. It offers critical insights to help companies decide how best to allocate their cloud budget.

Consumption Pricing Model: Pros and Cons

The rise of the consumption-based pricing model in cloud computing and SaaS subscription products enables product usage and businesses to scale usage with demand. However, while it offers flexibility, it also comes with some challenges. In this section, as stated in Figure 1-6, we break down the pros and cons to help you determine if this pricing model is the right fit for your business.

CHAPTER 1 FINOPS FOR SNOWFLAKE

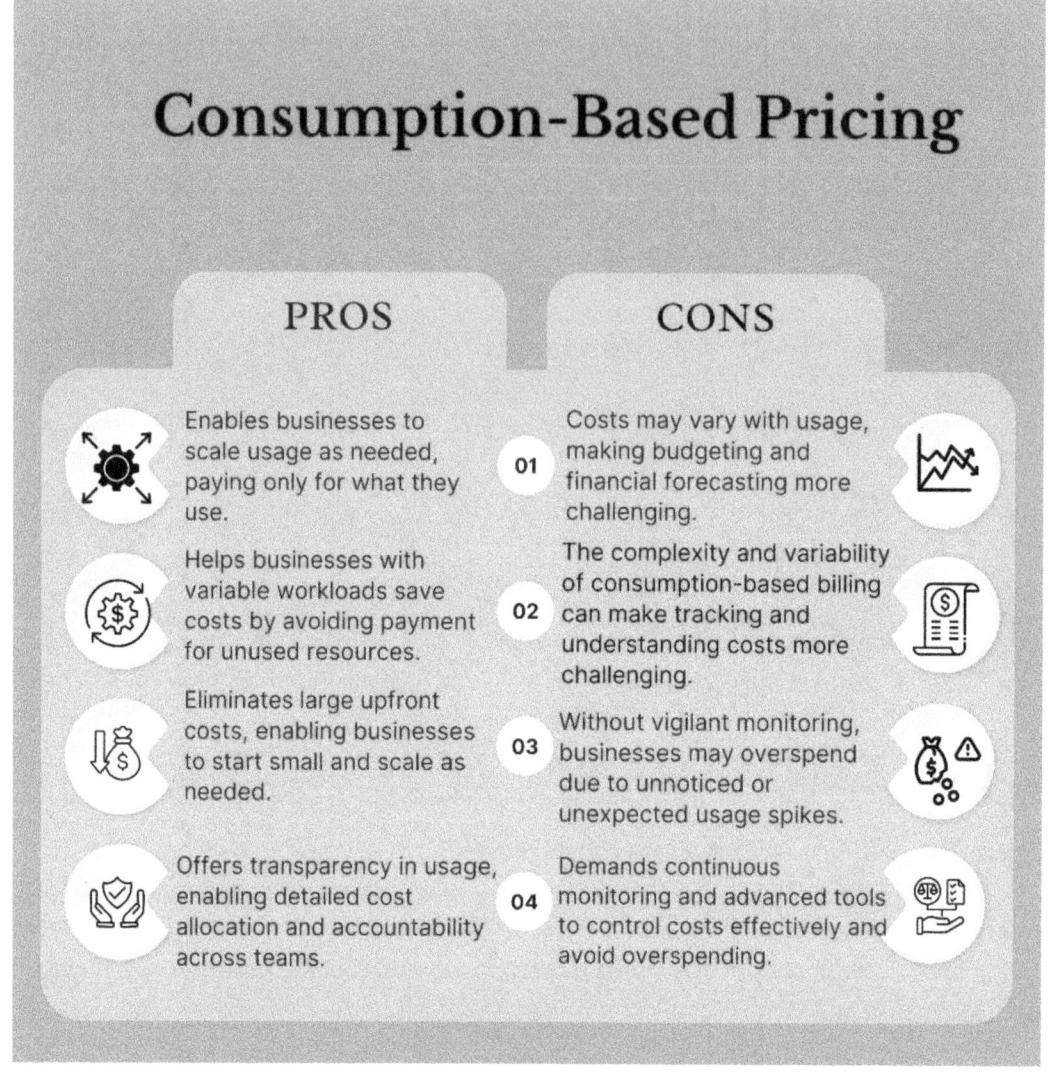

Figure 1-6. The pros and cons of consumption-based pricing

As companies migrate their services to on-demand cloud, they generally struggle with the same issue: lack of transparency into cost dynamics and understanding how they can influence and manage spend. Cost unawareness translates into wasted resources and budget overruns. These challenges prompt organizations to consider steps that will keep cloud spending in line with business objectives. The next section examines the critical areas of FinOps for the consumption model.

Challenges Without FinOps

Managing consumption-based cloud services without a dedicated FinOps strategy can lead to significant challenges that hinder effective financial oversight. Here are a few critical issues that organizations may face:

- **Unpredictable costs:** The on-demand cloud option provides a usage-based pricing model. It can be challenging to comprehend how money is spent and identify areas with potential for optimization.

- **Lack of cost awareness:** Most organizations and their employees are unaware of cloud usage costs or how their behavior can lead to unnecessary expenses. They lack the right tools or vision to gain visibility into cloud expenses.

- **Inadequate cost controls:** Automated management of resource usage might be missing, leading to wasted and underutilized resources.

- **Difficult to align cloud spend with business goals:** A flexible pricing model means that aligning cloud spending with business goals is critical to getting the most value from these investments. Organizations often struggle to connect cloud costs back to business outcomes, making it challenging to justify expenses and make data-driven decisions.

How FinOps Addresses Those Challenges

FinOps is designed to fit on top of the complexities of the variable pricing model by adding structure to cloud financial management. This framework is constructed on a series of core principles:

- **Collaboration and communication:** FinOps fosters a culture of collaboration among finance, technology, and business teams to align on common cost management goals. Open and transparent communication creates an environment where cost sensitivity is the norm rather than the exception, fostering accountability.

- **Data-driven decision-making:** Organizations should focus on using data-driven insights to drive decision-making. This information helps quickly identify overcharged intervals and measurements, determining whether costs are above expectations and aiding in provisioning decisions.

- **Persistent monitoring and refinement:** FinOps emphasizes practices that track cloud spend and waste, responding to lower spending. Efficiency and frictionless operations are always on the target list of organizations.

- **Flexibility and scalability:** FinOps guides organizations to adapt their financial processes as cloud usage matures. This is especially important for consumption-based models, which can vary depending on usage and costs.

- **Education and training:** Providing guidance to teams on optimizing cloud costs is part of FinOps. Indirect benefits come from employees deciding to conserve and cut back on wasteful activities, given a basic understanding of cost drivers and best practices.

Before we delve further into the key benefits of FinOps principles, it would be useful to revise the thoughts touched upon in the previous sections. In a nutshell, to sum up the key takeaways so far, it is crucial to define clear boundaries and set strict expectations among multiple stakeholders using the platform while stressing the need to get well-versed with platform usage and its best practices. The next section takes a deeper look into these crucial principles and their proven benefits to derive maximum business value and wise spending.

Advantages of FinOps Implementation

Stateless general-purpose services embody the essence of the FinOps model, which is nothing but characteristics that only exist in a consumption-based framework that offers:

> **Increased cloud usage:** Monthly cloud spend for organizations can be decreased by significant margins with appropriate controls over cloud usage.

Principled financial planning: Improved accuracy in forecasting and budgeting helps organizations avoid surprises as far as cloud billing usage is concerned. This aids in redistribution of resources and controls the unpredictability of funds.

Increased business agility: FinOps can utilize FinOps practices to quickly adjust the cloud usage based on changing market demands, which helps organizations stay ahead of their competition. Instead of rigid long-term plans, FinOps promotes an agile approach, allowing for iterative planning and faster response times.

Improved accountability and transparency: FinOps adopts a culture of financial accountability, where everyone in an organization is aligned on cost optimization goals.

So far, we've covered the factors that influence consumption pricing, and the key areas that companies should focus on to manage costs. The next section dives deeper into how the FinOps framework accelerates the cost-to-business ratio for Snowflake.

FinOps for Snowflake

FinOps for Snowflake emphasizes three critical dimensions necessary for optimizing cost management and presents effective best practices designed to drastically reduce expenditures on compute, storage, and data transfer. By adopting proactive management strategies and committing to continuous improvement, organizations can achieve substantial long-term cost savings and maximize resource utilization. Each of these dimensions is crucial for effective financial management in a consumption-based model, where cost control and optimization are essential for maintaining optimal performance within budgetary constraints.

The three main pillars of the FinOps Framework, displayed in Figure 1-7, establish a model that provides significant operational efficiency and allow organizations to take control of their costs. These pillars encompass strategies for forward-looking cost tracking, continual optimization efforts, and active cooperation between financial and technical teams to align with strategic corporate initiatives. By targeting these critical components, organizations can build a strong and transparent path to Snowflake cost management, promoting growing efficiencies and delivering improved financial results that keep them ahead of the competition.

CHAPTER 1 FINOPS FOR SNOWFLAKE

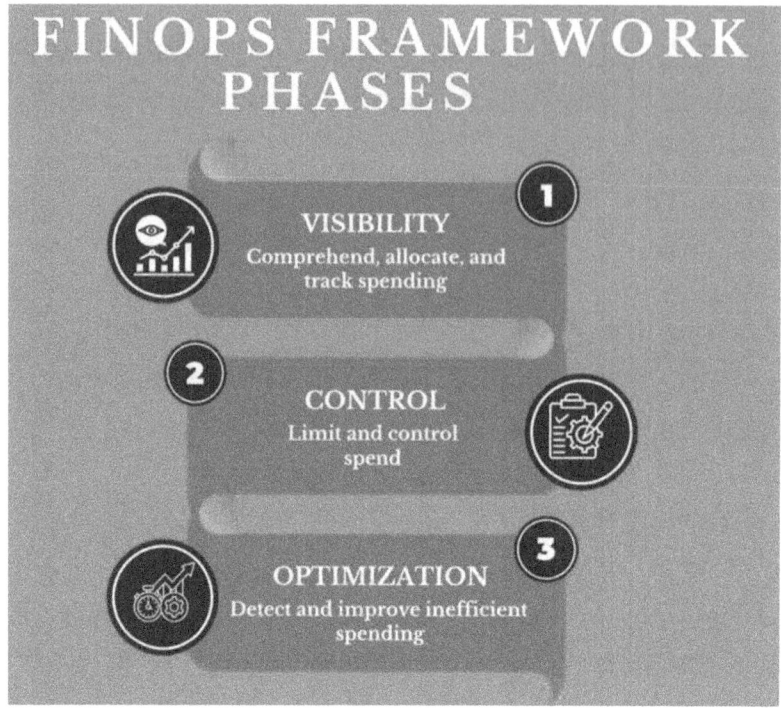

Figure 1-7. FinOps framework phases

Visibility and Transparency

- Establish visibility into your Snowflake usage trend and the underlying cost drivers.

- Use granular cost attribution and tagging to identify the source of costs.

- Track expenses and improve efficiency.

Control and Budgeting

- Forecast and budget strategies help to find gaps in variable spending.

- FinOps uses a rule-based prediction model where it takes past use and predicts the future cost.

- Establish budgets that align with your financial objectives, ensuring better control over spending.

Optimize Resource Utilization

- Techniques such as right-sizing compute and defining data lifecycle policies help organizations pay only for what they need.

- Automation is essential to manage Snowflake resources in response to real-time usage, which helps prevent runaway costs.

By focusing on these three key pillars, organizations can better manage cloud spending in Snowflake. Chapter 2 dives deeper into each pillar, exploring best practices and strategies for successful cost governance. From here, we take a closer look at the FinOps strategy, examining the essential elements for effective cost management.

A Comprehensive Approach to the FinOps Strategy

To realize the full potential of Snowflake FinOps, it is imperative to take a holistic approach. This approach is framed around four pillars—Strategy, People, Processes, and Technology—as depicted in Figure 1-8. By focusing on these components, you can unlock a complete FinOps approach that balances operational excellence and business value execution.

These elements, when considered together, ensure that FinOps initiatives within an organization are easily integrated, sustainable, and aligned with the broader objectives of the organization.

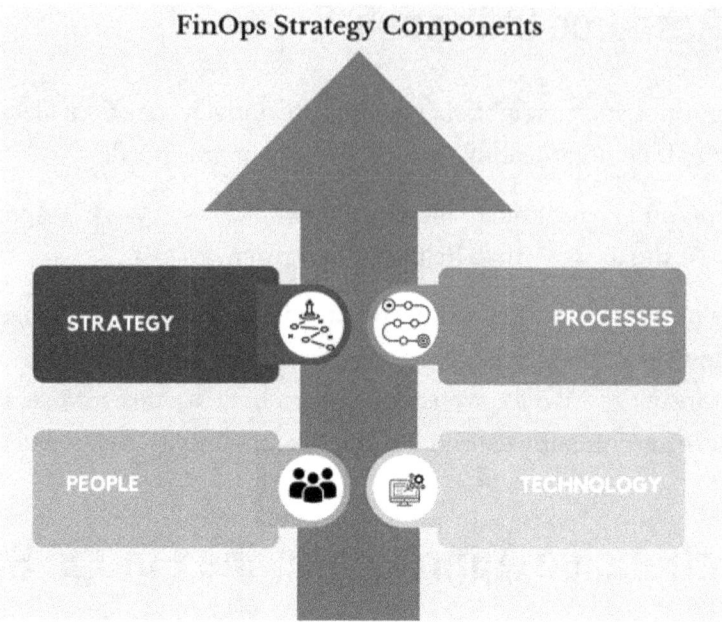

Figure 1-8. FinOps strategy components

Strategy involves aligning Snowflake usage with broad data goals and establishing key performance indicators (KPIs) to measure the impact of FinOps initiatives.

People focus on encouraging a culture of cost awareness by creating dedicated FinOps teams and equipping them with the knowledge and skills to optimize cloud spending and best practices.

Processes emphasize the importance of standardizing cost allocation, budgeting, forecasting, continuous feedback loops, and implementing chargeback models for accurate cost attribution.

Technology utilizes Snowflake's built-in powerful tools like Snowsight, Account Usage, Organization Usage views, and budgeting features to streamline cost management processes.

In this section, we help users integrate these components into the FinOps framework, enabling them to run an iterative process to manage and optimize Snowflake spend, as depicted in Figure 1-9.

CHAPTER 1 FINOPS FOR SNOWFLAKE

Figure 1-9. Snowflake FinOps strategy

Investing in Skills

- **Importance of FinOps skills:** Recognize that FinOps is a distinct skill that offers different objectives from traditional software development and DevOps roles.

- **Talent development/recruiting:** Organizations preparing to build a FinOps capability should invest in recruiting and upskilling existing employees.

- **Certification achievements:** Embed a culture of continuous learning by motivating more team members to take FinOps certifications.

Understanding User Needs

- **Tailored approaches:** Knowing your users allows you to customize your cloud services according to their specific requirements.

- **Organization culture:** Bring FinOps aligned with your organization's culture and values.

KPIs (Key Performance Indicators)

- **Value-driven KPIs:** Driven by metrics that influence overall financial health and operational efficiency rather than being completely cost-based.

- **Celebrate successes:** Track and celebrate milestones through KPIs. This not only motivates teams but also reinforces the importance of cost management.

Culture of Spend Awareness

- **Embedded awareness:** Change the culture to talk about cloud spend.

- **Incorporate it into conversations:** Normalize owning financial accountability through regular discussions to showcase and address cloud spending.

- **Steps to enhance cost awareness:** Organize initiatives (such as workshops and meetups) to boost cost awareness and share best practices.

Empowering Teams

- **Selecting the right tools:** Be cautious while selecting tools and choose the one that can solve your particular operational demand instead of chasing shiny tools.
- **Operationalizing change:** Simplify the adoption of new processes and workflows while providing them with the necessary resources to adapt to new ways of working.

Tracking What Counts

- **Metric momentum:** Focus on a few key metrics that directly impact the cloud spend bottom line.
- **Celebrate successes:** Highlight achievements with metrics to build momentum and reinforce the importance of financial accountability.

Continuous Improvement

- **Feedback loops:** Establish mechanisms for periodic feedback and evaluate the improvements derived from this feedback.
- **Maturity model progress:** Always track and log progress to ensure continuous improvement.

Gamification

- **Strategies to motivate:** Motivate employees and gain a sense of teamwork through gamification. This can turn efficient spending into a point of pride.
- **Participation metrics:** Measure success on adoption and usage to know how many teams are actively playing the cost-saving gamified challenges.

While these strategies provide an effective foundation for controlling costs, it's equally important to involve stakeholders across the organization. A key factor in long-term success is involving stakeholders from finance and engineering to operations and leadership. This crucial practice is the key underlying principle for accomplishing cost optimization with high business value.

The next section shifts focus to a shared accountability model and explores the challenges of implementing it.

Shared Accountability Model Challenges

The shared accountability model requires a combined effort from all teams involved. Financial accountability is shared, meaning every team has a role in driving cost optimization and overcoming the challenges that come with it. Since finance, technology, and business own the entire operation, collaboration and clear communication are critical. Any misalignment can lead to inefficiencies and cost overruns.

In the shared accountability model, there is no single team responsible for success or failure. Instead, all teams must work together to ensure that the infrastructure is used optimally, costs are managed effectively, and financial goals are met. This shared responsibility means that each team has a direct impact on cost control and operational efficiency. However, it also introduces several challenges.

Finance needs to:

- **Cloud forecast spending:** There is a challenge in forecasting accurate monthly and yearly costs, which can put a burden on the budget.

- **Define budget based on predictions:** Inflexible budgets can limit the ability to respond to unexpected changes and opportunities.

- **Maintain changes to the budget as needed:** Reacting to unpredictable cost fluctuations can strain resources and hinder effective financial planning.

Technology needs to:

- **Be accountable for overall spending on their platform:** Lack of ownership may lead to unchecked expenditures.

- **Understand cost inefficiencies:** Identifying cost inefficiencies without visibility is quite complex.

- **Ensure transparency on "cost spent" and "where are my problems":** Communication blackouts can mask spending visibility.

- **Utilize unit costs to project and budget spend:** Variability in resource usage makes it difficult to project costs.

Business needs to:

- **Control their budget and prepare for overruns:** Whenever there are some unexpected costs, a variance will ripple through the budget.

- **Manage cost optimization vs. business goals:** It is often difficult to balance the cost optimization with the wider business goals when priorities conflict.

Figure 1-10 shows some of the key strategic components required to overcome these challenges and achieve the shared accountability model.

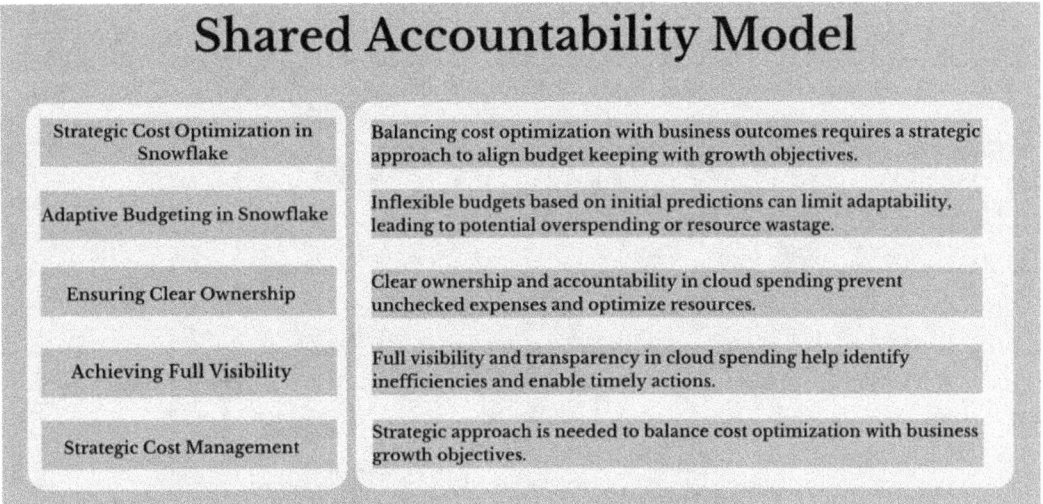

Figure 1-10. The shared accountability model

Now that we have established accountability among all stakeholders, the next section moves to the cost optimization framework, which is designed to help effectively manage Snowflake costs.

The Cost Optimization Framework for Snowflake

Managing costs and optimizing performance are among the most critical challenges in cloud data warehousing in today's data-driven world. Investing in a new approach to cost management presents an opportunity to significantly improve your ROI on Snowflake. This section explores the basics of FinOps for Snowflake and the strategies needed to optimize your cloud spending. Given the rapid increase in cloud consumption, it is crucial to manage and control your resources effectively. Lack of proper management can lead to higher costs and lost productivity.

As illustrated in Figure 1-11, this chapter delves into the core principles of FinOps for Snowflake—a strategic approach to help organizations gain better control over their cloud consumption. The focus is specifically on proactive strategies to optimize spending, streamline operations, define AI usage, and enhance collaboration across teams. This approach empowers your team to make smarter decisions and sustain long-term success in the cloud.

Figure 1-11. Optimizing cost management and efficiency

Incident Fire Fighting

Quickly responding to unplanned incidents ensures that related costs do not increase sharply. Establish an operational process to assemble a team that can handle incidents promptly to provide a timely and efficient response. Set up alerts for anomalies and budget thresholds. Leverage `QUERY_HISTORY` view from Snowflake to rapidly diagnose issues. Conduct regular drills to simulate incident scenarios and improve response times. Managing outliers in resource usage is a key part of cost control.

Cost Tracking using Query Tag

Query Tag provides users with the capability to map their queries to various attributes, such as projects, cost centers, and specific teams. By analyzing queries that experience high utilization, teams can identify opportunities for cost reduction and efficiency enhancement. Such transparency facilitates budgetary planning and enables teams to utilize resources in a more effective manner.

Map Owners Accurately

Strategically mapping Snowflake usage with application and domain owners creates an accountability atmosphere in controlling cost. Along with query and object tagging, organizational metadata helps roll up the usage report at different levels in the company. This inclusive measure of visibility allows organizations to ensure optimized cloud costs.

Get Everyone to Help

FinOps is a team effort. To build a culture of cost awareness, organizations should bring together teams from different areas to work with data engineering, analytics, and finance. Regular training should be provided to help teams understand how their actions affect costs. You can promote teamwork through joint efforts to reduce costs. Set up feedback loops so everyone can share insights and learn from each other.

Use Automation to Detect Outliers

Automated outlier detection tools can significantly reduce manual effort in monitoring workload and performance. Integrating Snowflake with third-party monitoring tools will be helpful to receive real-time alerts about abnormal spending patterns. You can proactively address problems utilizing Snowflake's AI and ML capabilities to identify and address potential issues before they escalate.

Use Automation to Create Self-Service Queries

Data engineers often deal with a high volume of complex query requests. By providing self-service capabilities, you can reduce dependency on platform engineers. It also empowers users to identify and optimize expensive queries independently. Improved accountability through dashboards that flag the queries promotes responsible usage of resources. Cost management empowers users to make proactive adjustments to their queries.

System Level Optimization

To optimize Snowflake at the system level, a deep knowledge of architecture and usage patterns are essential. Analyze application workload and storage usage regularly and identify optimization opportunities. Develop system optimization strategies like adjusting warehouse resources based on workload demands, using query acceleration services, and auto-suspending/resuming idle warehouses to reduce costs. Snowflake resource monitor and Snowsight help track usage metrics and proactively adjust configurations to ensure optimal resource allocation and cost efficiency.

Evolution of the Snowflake System

As Snowflake continues to evolve to serve evolving data needs, companies need to adjust their FinOps practices to stay on top. To remain competitive with the latest innovations and stay updated on new features, companies must participate in Snowflake community forums and attend webinars. Contributing to these community forums is another way to stay up on the latest features and best practices.

Data Operations

Using operational data and financial metrics, companies can uncover valuable insights that drive informed decision-making. Seamlessly integrated dashboards provide a holistic view of both operational efficiency and cost management, enabling stakeholders to make strategic decisions that align business goals with resource utilization.

Fine-Tune AI-Based Suboptimal Patterns

Utilize AI and ML capabilities to study historical performance data of query execution in order to discover non-optimal patterns in SQL and track the underlying reasons for misuse of cost-incurring resources. Retrain AI models to use patterns that change over time and continue optimizing.

To sum up the key takeaways so far, to manage costs effectively, begin by establishing clear processes and empowering your team to take ownership. Ensure that they're responsible for learning the technology, understanding the requirements, and optimizing the costs. Focus on identifying recurring incidents, analyzing query patterns, and applying automation, alerts, and other cost-saving techniques. Keep exploring new processes that contribute to cost reduction. Use the Snowflake FinOps model as a starting point and feel free to customize and adapt it to best suit your needs over time. This approach will help you build a sustainable framework for long-term cost management and optimization.

Summary

This chapter highlighted the importance of cloud cost-management best practices in today's consumption-based economy, emphasizing the value of adopting a FinOps approach. At its core, FinOps is about setting clear boundaries, establishing strong expectations, and ensuring that the right stakeholders take ownership of their consumption. Effective management comes from using best practices, consistently reviewing usage, and staying mindful of spending—all the while ensuring that every dollar spent delivers maximum value for the business.

While cost savings are important, the bigger picture is about justifying those costs through the value they bring to the business. FinOps practices act as a guiding post for accomplishing financial goals with operational outcomes.

CHAPTER 1　FINOPS FOR SNOWFLAKE

The FinOps model is very broad, and this book does not cover all its aspects in detail. The objective is to present impactful practices that will help you optimize costs effectively. In the upcoming chapters, we dive into cost optimization strategies, supported by real-world use cases, to help you achieve your goals.

CHAPTER 2

FinOps Adoption for Snowflake

Overview

This chapter explores how organizations can adopt the FinOps framework to effectively manage and optimize Snowflake usage and costs. By aligning financial operations with the architecture and capabilities of Snowflake, teams can gain deep visibility into resource consumption, enforce cost controls, and drive continuous optimization. The FinOps journey is structured around three progressive phases—Visibility and Transparency, Control and Budgeting, and Optimization.

By using native tools in combination with governance and automation, organizations can make informed decisions, reduce waste, and align cloud spending with business value. This chapter provides a practical roadmap for embedding FinOps principles into Snowflake environments, ensuring scalable and sustainable cost management through:

- Using native tools and tagging for granular cost visibility
- Establishing governance and consistent cost attribution practices
- Implementing budgets, quotas, and automated alerts for cost control
- Automating idle resource suspension and dynamic scaling
- Optimizing performance using caching, clustering, and materialized views
- Promoting shared accountability through chargeback and showback models

To achieve this vision, organizations must adopt a structured approach that integrates financial operations with the technical capabilities of Snowflake. The next section introduces a purpose-built FinOps framework for Snowflake, outlining the foundational practices that support cost efficiency, governance, and alignment with business goals.

FinOps Framework for Snowflake

Adopting FinOps in a Snowflake environment requires more than just monitoring costs. It calls for a cultural and operational shift that embeds financial accountability into the daily workflows of the engineering, finance, and business teams. This section introduces a structured FinOps framework specifically designed for Snowflake, enabling organizations to manage cloud spend with precision and purpose.

As illustrated in Figure 2-1, the framework is built around three core phases—Visibility, Control, and Optimization. These phases represent a progressive journey toward financial maturity. Each phase introduces specific tools, practices, and governance models that help teams understand their data consumption, enforce cost controls, and continuously improve performance and value.

- **The Visibility and Transparency phase** focuses on surfacing cost drivers and usage patterns using Snowflake's native tools such as account usage views, tagging, and third-party integrations. This phase lays the foundation for transparency and informed decision-making.

- **The Control and Budgeting phase** introduces mechanisms to enforce budgets, automate governance, and implement policies that prevent overspending. It includes tools like resource monitors, quotas, and chargeback or showback models to drive accountability.

- **The Optimization phase** translates insights and controls into action. This includes tuning queries, leveraging caching and materialized views, and dynamically scaling compute to maximize efficiency and business value.

Customizing this framework involves aligning it with the organization's goals, usage patterns, and team structures. It requires collaboration across departments, a commitment to data-driven decision-making, and the implementation of automation and governance mechanisms that scale with growth.

Snowflake's native capabilities—such as automated scaling, tagging, and cost monitoring—integrate seamlessly into this framework. By embedding FinOps into the Snowflake ecosystem, organizations can move from reactive cost tracking to proactive financial stewardship. This ensures that cloud investments deliver measurable value while fostering a culture of shared responsibility and continuous improvement.

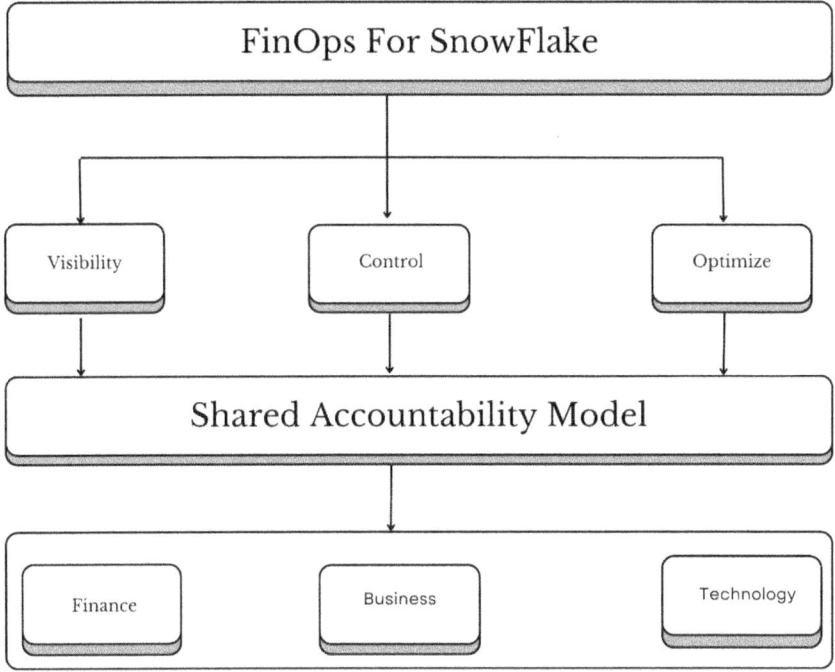

Figure 2-1. *FinOps framework for Snowflake*

CHAPTER 2 FINOPS ADOPTION FOR SNOWFLAKE

With a solid understanding of the FinOps framework established, the journey begins with the first and most foundational phase: Visibility and Transparency. This phase sets the stage for all subsequent efforts by delivering the insights necessary to understand, manage, and optimize Snowflake costs effectively.

Visibility and Transparency Phase

The Visibility and Transparency phase is the foundation of effective FinOps adoption in Snowflake. Without a clear understanding of how resources are consumed and where costs originate, organizations cannot manage or optimize their Snowflake environment effectively. This phase focuses on surfacing cost drivers, identifying usage patterns, and establishing the groundwork for accountability and informed decision-making.

As illustrated in Figure 2-2, this phase is composed of several key components that work together to provide a comprehensive view of Snowflake usage and spending. By leveraging Snowflake's native tools, integrating third-party solutions, and implementing governance practices, organizations can build a transparent cost structure that supports both operational efficiency and financial control.

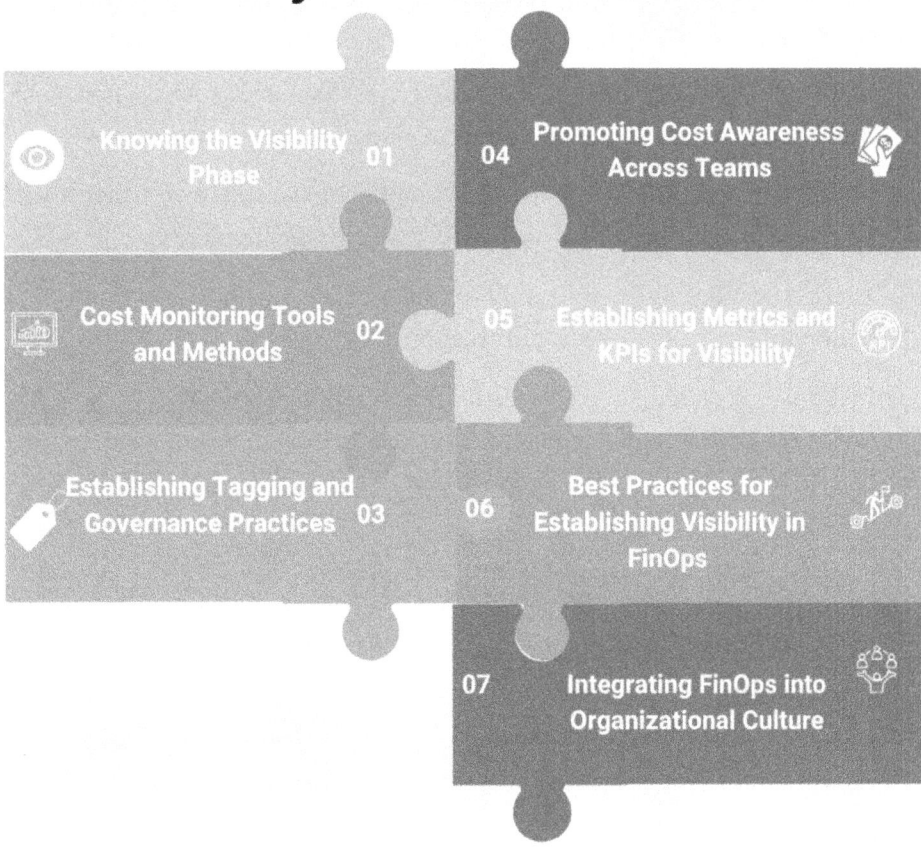

Figure 2-2. Visibility and Transparency phase for Snowflake

The following sections break down the core practices of the Visibility and Transparency phase and their impact on FinOps success in Snowflake.

Understanding the Visibility and Transparency Phase

At the heart of this phase is a critical question: Where is our money going? By aggregating and analyzing cost and usage data, organizations can uncover the true sources of Snowflake expenses and begin managing them with greater precision.

This phase emphasizes the importance of understanding the Snowflake pricing model, which includes compute, storage, and data transfer costs. It also involves identifying key cost drivers such as warehouse use, long-term storage, and cross-region data movement.

The main objectives of the Visibility and Transparency phase are:

- Understanding the cost structure by gaining clarity on how Snowflake charges for various services.
- Identifying cost drivers that contribute most to overall spend, such as inefficient queries or underutilized resources.
- Enabling insightful analysis to detect inefficiencies, uncover usage trends, and evaluate spending across teams or departments.
- Driving accountability by equipping stakeholders with transparent, actionable insights that support responsible use.

Organizations that establish strong visibility are better equipped to make informed decisions, optimize their Snowflake environments, and build a foundation for long-term cost control.

Cost Monitoring Tools and Methods

Effective cost monitoring is essential for managing Snowflake usage and ensuring financial accountability. This involves not only tracking consumption but also establishing a system that enables proactive decision-making and cost control.

Snowflake provides several native tools that offer valuable insights into usage patterns:

- **Account usage views:** Tables such as QUERY_HISTORY, WAREHOUSE_METERING_HISTORY, and STORAGE_USAGE_HISTORY allow teams to monitor query performance, warehouse activity, and storage consumption in detail.
- **Resource monitors:** These enable administrators to set credit thresholds and receive alerts when usage exceeds defined limits, helping to prevent unexpected cost overruns.

While these native tools form a strong foundation, organizations often benefit from extending their monitoring capabilities through additional methods:

- **Third-party FinOps platforms:** Tools available in the marketplace can aggregate data from multiple sources, offering centralized visibility and more granular insights into Snowflake usage and costs.

- **Custom dashboards:** Visualization tools such as Tableau, Power BI, and Looker can be used to build dashboards that track usage trends, highlight anomalies, and support real-time monitoring.

- **Automated reporting:** Scheduled reports can be distributed to stakeholders on a weekly or monthly basis, ensuring that cost data is consistently reviewed and acted upon.

Combining Snowflake's built-in capabilities with structured monitoring practices enables organizations to maintain transparency, support cost accountability, and take timely action to optimize resource usage.

Establishing Tagging and Governance Practices

Tagging and governance are foundational to achieving visibility and accountability in Snowflake cost management. Without a consistent and enforced tagging strategy, it becomes difficult to attribute costs accurately, monitor usage by business unit, or implement effective controls.

A well-defined tagging framework enables organizations to track resource consumption by department, project, environment, and cost center. These tags serve as metadata that supports cost allocation, reporting, and optimization efforts.

Key practices for effective tagging and governance include:

- **Implementing tagging standards:** Define a consistent tagging schema that includes attributes such as department, project name, environment (e.g., development, test, production), and cost center. Apply these tags across all Snowflake objects, including warehouses, databases, schemas, and roles.

- **Establishing a governance model:** Form a cross-functional governance team responsible for maintaining tagging standards, enforcing cost allocation policies, and overseeing FinOps practices. This team should also define roles and responsibilities for managing cost-related configurations.

- **Utilizing native cost allocation features:** Leverage Snowflake's built-in cost visibility tools, such as cost center columns and account usage views, to analyze workload impact and ensure accurate attribution. Regularly review and update tagging structures to reflect organizational changes and evolving project needs.

- **Enforcing tagging compliance:** Use automation and policy enforcement to ensure that all new resources are tagged appropriately. Prevent the creation of untagged or misconfigured resources by integrating tagging checks into provisioning workflows.

By embedding tagging and governance into Snowflake operations, organizations can ensure traceability, promote accountability, and lay the groundwork for scalable and transparent cost management.

Promoting Cost Awareness Across Teams

Creating a culture of cost awareness is essential for sustainable Snowflake usage. When teams understand how their actions impact cloud spending, they are more likely to take ownership and contribute to cost optimization efforts. Promoting this awareness requires a combination of communication, education, and accountability mechanisms.

Key strategies to promote cost awareness include:

- **Regular cost review meetings:** Schedule recurring sessions with finance, engineering, and data teams to review Snowflake usage and spending reports. These meetings help identify optimization opportunities and encourage teams to manage their own budgets and resource consumption.

- **Chargeback and showback models:** Implement chargeback to allocate actual Snowflake costs to departments or projects, or use showback to provide visibility without direct billing. Both models foster accountability and align with FinOps principles by making teams more conscious of their usage.

- **Training and knowledge sharing:** Offer targeted training sessions on Snowflake cost monitoring, usage tracking, and optimization techniques. Share best practices, real-world examples, and success stories to build engagement and empower teams to participate actively in FinOps initiatives.

By embedding cost awareness into daily operations and team responsibilities, organizations can drive more responsible usage and ensure that Snowflake investments deliver measurable business value.

Establishing Metrics and KPIs for Visibility

Defining clear metrics and key performance indicators (KPIs) is essential for measuring the effectiveness of the Visibility phase in a FinOps strategy. These indicators help organizations track progress, identify inefficiencies, and demonstrate the value of improved cost transparency across Snowflake environments.

Metrics should be aligned with business goals and provide actionable insights that support both operational and financial decision-making. They also serve as a foundation for accountability, enabling teams to monitor their impact and adjust behaviors accordingly.

Key metrics to monitor include:

- **Total Snowflake spend:** Track overall spending trends to identify spikes, anomalies, or seasonal patterns that may require investigation or adjustment.

- **Resource consumption by business unit or project:** Measure compute and storage usage across departments or initiatives to support accurate cost allocation and budgeting.

- **Warehouse activity and utilization:** Monitor warehouse usage to detect idle or underutilized resources, which can be resized or consolidated to improve efficiency.

- **Tagging coverage and compliance:** Evaluate the percentage of resources that are properly tagged to ensure traceability and support cost attribution.

- **Query performance and efficiency:** Analyze query execution times and scan volumes to identify optimization opportunities that reduce compute costs.

- **A FinOps scorecard:** Develop a centralized scorecard that consolidates key metrics, KPIs, and usage trends. This scorecard should be shared regularly with stakeholders to communicate progress, highlight areas for improvement, and reinforce a culture of financial accountability.

By establishing and tracking meaningful metrics, organizations can ensure that visibility efforts translate into measurable outcomes and continuous improvement.

Best Practices for Establishing Visibility in FinOps for Snowflake

Establishing effective visibility in Snowflake requires more than just enabling tools. It demands a strategic, phased approach grounded in proven practices. These best practices help organizations gain actionable insights, minimize implementation risks, and build a strong foundation for FinOps maturity.

- **Start with a pilot project:** Begin by applying visibility practices such as tagging, reporting, and monitoring to a single team or department. This controlled rollout allows for early feedback, process refinement, and smoother scaling across the organization.

- **Launch a FinOps center of excellence (CoE):** Form a cross-functional team that includes representatives from finance, engineering, procurement, and operations. The CoE should lead visibility initiatives, standardize processes, and serve as a central hub for FinOps governance and knowledge sharing.

- **Conduct regular usage audits:** Perform periodic audits of Snowflake resources to identify unused warehouses, orphaned data, or inefficient configurations. These audits help maintain real-time visibility and prevent unnecessary spending.

- **Enable transparent communication:** Establish clear communication channels between finance, IT, and engineering teams. Sharing cost reports, usage trends, and optimization insights fosters collaboration and ensures that visibility efforts are aligned with business goals.

By following these practices, organizations can build a scalable and resilient visibility framework that supports long-term cost management and operational efficiency in Snowflake.

Integrating FinOps into the Organizational Culture

Achieving true visibility in Snowflake cost management goes beyond tools and processes. It requires a cultural shift. Embedding FinOps into the organization means aligning values, behaviors, and accountability with financial transparency and operational efficiency.

- **Executive sponsorship and buy-in:** Secure leadership support to give FinOps initiatives enterprise-wide significance. When executives champion cost visibility, it sets a tone from the top that encourages teams to prioritize financial accountability.

- **Foster a data-driven mindset:** Promote a culture where decisions are guided by cost and usage data. Encouraging teams to regularly reference metrics builds long-term awareness of Snowflake consumption and its financial impact.

- **Empower cost champions:** Appoint FinOps advocates within each team to oversee Snowflake use, enforce tagging standards, and share insights. These champions help sustain visibility and drive continuous improvement across the organization.

> **Note** Visibility and transparency practices are explored in greater depth in Chapter 3, with particular emphasis on customized reporting and pricing awareness.

With this foundation in place, the next step is to move from awareness to action through structured control and budgeting mechanisms.

Control and Budgeting Phase

Following the Visibility phase, the Control and Budgeting phase marks a critical transition from insight to action. This phase focuses on applying governance policies, implementing budget controls, and leveraging automation to manage Snowflake costs proactively. It ensures that the transparency gained earlier is translated into tangible financial discipline and operational efficiency.

During this phase, organizations establish a structured control plane that includes policy enforcement, automated cost-saving mechanisms, and accountability frameworks such as chargeback and showback models. These practices help teams stay within budget, avoid unexpected expenses, and align Snowflake usage with business priorities.

As illustrated in Figure 2-3, this stage introduces the foundational elements required to maintain financial control while supporting scalable and efficient Snowflake operations. This phase also fosters a collaborative ecosystem where cost accountability is shared across teams, enabling continuous feedback and improvement.

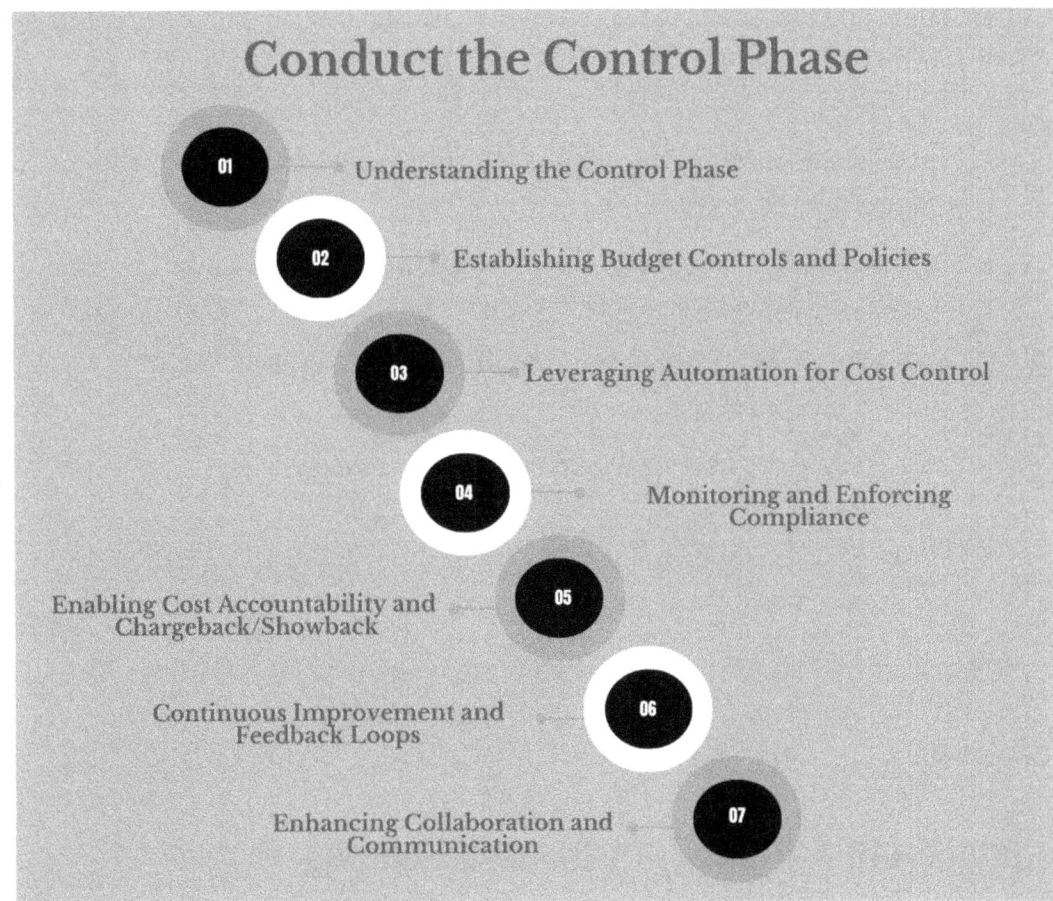

Figure 2-3. Conducting the Control and Budgeting phase

Understanding the Control and Budgeting Phase

The Control and Budgeting phase is where FinOps practices shift from observation to execution. Building on the insights gained during the Visibility and Transparency phase, this stage focuses on implementing governance, automation, and policy enforcement to manage Snowflake costs in a structured and proactive way.

This phase introduces mechanisms to define usage policies, enforce budgets, and automate responses to cost anomalies. It also emphasizes continuous monitoring and real-time adjustments to prevent overspending and ensure that Snowflake resources are used efficiently and in alignment with business goals.

The Control and Budgeting phase transforms cost management into a repeatable and scalable discipline that supports financial accountability across teams.

Key objectives of the Control and Budgeting phase include:

- Establishing usage policies across all Snowflake accounts to standardize cost governance.

- Automating cost controls to cover at least 80 percent of daily operations, reducing manual intervention.

- Maintaining budget compliance within a 5 to 10 percent variance from planned spending.

- Achieving 100 percent tagging compliance for accurate cost attribution and reporting.

- Enabling chargeback or showback models for all business units to promote ownership and accountability.

By meeting these objectives, organizations can move from reactive cost tracking to proactive financial management, setting the stage for long-term efficiency and optimization.

Establishing Budget Controls and Policies

Establishing effective budget controls is essential for maintaining financial discipline in Snowflake environments. This involves setting clear spending limits, enforcing governance policies, and using automation to monitor and manage usage in real time.

To begin with, organizations should define budgets at the account and project levels. These budgets should be aligned with business objectives and reviewed regularly. For example:

- Set monthly or quarterly budgets for each Snowflake account.

- Allocate project-level budgets with a variance tolerance of no more than 10 percent.

- Configure automated alerts at 75 percent, 90 percent, and 100 percent of budget thresholds to enable timely intervention.

Governance policies should be implemented to guide how Snowflake resources are provisioned and consumed. These policies may include:

- Restricting access to high-cost features unless explicitly approved.
- Enforcing tagging standards to ensure that 100 percent of resources are properly attributed.
- Limiting the creation of large or untagged virtual warehouses.

Resource quotas and usage limits are also critical. Organizations should:

- Apply compute and storage quotas to prevent overuse.
- Review and adjust quotas quarterly based on usage trends.
- Ensure that at least 95 percent of active resources fall within defined quota boundaries.

By combining budget controls with governance and automation, organizations can create a stable and predictable Snowflake cost environment. These practices help teams stay within financial targets while supporting scalable and efficient operations.

Leveraging Automation for Cost Control

Automation is a critical enabler of cost efficiency in Snowflake environments. By automating routine tasks, organizations can reduce manual overhead, respond to usage patterns in real time, and enforce financial governance at scale.

Automation should be applied across key operational areas to ensure consistent and proactive cost control. These include resource optimization, smart scaling, cost tracking, and compliance monitoring.

Key areas of automation include:

- **Resource optimization:** Automatically suspend inactive warehouses and delete unused data to reduce unnecessary compute and storage costs. Organizations should aim to automate at least 90 percent of idle resource management tasks.
- **Smart scaling:** Configure auto-scaling policies to match warehouse size with workload demand. This helps avoid overprovisioning and ensures that compute resources are used efficiently. Target a 20 to 30 percent reduction in average warehouse size through dynamic scaling.

- **Cost tracking and tagging:** Use automation to apply cost center, project, and team tags to all Snowflake resources. Enforce tagging compliance at the time of resource creation to maintain 100 percent attribution accuracy.

- **Monitoring and compliance enforcement:** Implement real-time tracking using Snowflake Account Usage and Snowsight. Build dashboards that visualize cost trends, policy adherence, and usage anomalies. Conduct automated audits weekly to ensure alignment with budgets and tagging policies.

- **Scorecards and accountability:** Generate automated scorecards that track compliance with cost policies and budget thresholds. Share these reports with stakeholders monthly to promote transparency and drive continuous improvement.

By embedding automation into cost control processes, organizations can ensure that Snowflake usage remains efficient, predictable, and aligned with financial goals. This approach not only reduces operational burden but also strengthens accountability across teams.

Monitoring and Enforcing Compliance

Effective cost control in Snowflake requires continuous monitoring and enforcement of financial policies. This ensures that teams stay within budget, adhere to governance standards, and take corrective action when needed. Monitoring and compliance are not one-time activities but ongoing processes that support accountability and operational discipline.

Organizations should implement real-time tracking tools to monitor Snowflake usage and spending as it occurs. Dashboards built with tools like Snowsight or third-party platforms can provide visibility into cost trends, policy adherence, and resource utilization. These dashboards should be reviewed regularly by both technical and financial stakeholders.

To maintain compliance, organizations should conduct scheduled audits of Snowflake environments. These audits should verify that spending aligns with budgets, tagging policies are enforced, and no unauthorized configurations exist. A monthly audit cadence is recommended, with ad hoc reviews triggered by anomalies or budget threshold breaches.

Scorecards are a valuable tool for measuring compliance. Scorecards should track key indicators such as:

- Percentage of resources with valid cost center tags (target: 100 percent)
- Number of budget alerts triggered and resolved within the same billing cycle
- Frequency of policy violations and time to remediation
- Percentage of teams reviewing cost dashboards at least once per week

Sharing these scorecards with stakeholders promotes transparency and reinforces shared responsibility. It also helps identify areas for improvement and supports a culture of continuous optimization.

By embedding monitoring and compliance into daily operations, organizations can ensure that Snowflake usage remains aligned with financial goals and governance standards.

Enabling Cost Accountability and Chargeback/Showback

Establishing cost accountability is a key outcome of the Control and Budgeting phase in the FinOps framework. By implementing chargeback and showback models, organizations can ensure that Snowflake costs are visible, traceable, and aligned with actual use. These models promote financial responsibility across teams and encourage more efficient consumption of cloud resources.

Chargeback involves directly allocating Snowflake costs to the teams or departments that generate them. This model fosters ownership and incentivizes teams to stay within budget. Showback, on the other hand, provides visibility into usage and costs without direct billing. It is often used as a transitional step toward full accountability.

To implement these models effectively, organizations should:

- Allocate 100 percent of Snowflake costs to business units, projects, or departments using automated tagging and cost attribution.
- Generate detailed cost reports that break down usage by team, environment, and workload. Reports should be shared at least monthly with stakeholders.

- Use visualization tools such as Power BI, Tableau, or Looker to create interactive dashboards that help teams understand their spending patterns and identify optimization opportunities.

- Appoint FinOps champions within each team to monitor Snowflake usage, interpret cost data, and lead cost-saving initiatives.

- Designate cost owners for each project or department who are responsible for staying within budget and adhering to financial policies.

By embedding chargeback and showback into Snowflake operations, organizations can create a culture of accountability that supports long-term financial sustainability and operational efficiency.

Continuous Improvement and Feedback Loops

Sustaining cost efficiency in Snowflake requires more than one-time optimizations. Continuous improvement and structured feedback loops are essential to evolving FinOps practices and adapting to changing business needs. This phase ensures that cost control becomes a dynamic, iterative process rather than a static set of policies.

Organizations should establish regular review cycles to assess the effectiveness of cost controls, automation, and governance. These reviews should include cross-functional stakeholders from finance, engineering, and operations. A quarterly cadence is recommended for strategic reviews, with monthly checkpoints for operational updates.

To support continuous improvement, teams should:

- Conduct post-implementation reviews of major FinOps initiatives to evaluate outcomes and identify lessons learned.

- Track and report on key performance indicators such as cost savings achieved, policy compliance rates, and tagging accuracy.

- Maintain a backlog of improvement opportunities based on audit findings, stakeholder feedback, and usage trends.

- Implement a feedback mechanism that allows teams to suggest enhancements to FinOps processes, tooling, or reporting.

Scorecards and dashboards should be updated regularly to reflect progress and highlight areas for refinement. These tools help maintain transparency and ensure that all teams remain aligned with the financial goals.

By embedding feedback loops into the FinOps lifecycle, organizations can continuously refine their Snowflake cost management strategy, respond to emerging challenges, and drive long-term value from their cloud investments.

Enhancing Collaboration and Communication

Collaboration and communication are essential to embedding FinOps into Snowflake operations. When teams work together and share insights, cost optimization becomes a shared responsibility rather than a siloed task.

- **Cross-functional FinOps meetings:** Hold regular meetings with finance, engineering, and product teams to review Snowflake usage, align on cost-saving goals, and share accountability for budget performance.

- **Centralized communication channels:** Use dedicated spaces in tools like Slack or Microsoft Teams to share updates, budget alerts, and best practices. These channels help maintain transparency and encourage real-time collaboration.

- **Training and knowledge sharing:** Provide ongoing education through workshops, quick guides, and recorded tutorials. Equip teams with the knowledge to interpret cost data and take informed action.

With governance and cost controls in place, organizations can now shift from managing spend to maximizing value.

Note Best practices for taking control of Snowflake spend are explored in greater depth, with a particular focus on resource monitoring, budgeting, and implementing chargeback and showback models, as detailed in Chapter 4.

The next phase, Optimization, focuses on improving performance and efficiency to ensure every Snowflake credit delivers measurable impact.

Optimization Phase

This phase, as illustrated in Figure 2-4, focuses on improving performance and reducing costs through various strategies, including query tuning, automatic clustering, materialized views, and caching. These strategies are essential for optimizing Snowflake usage and ensuring efficient resource management.

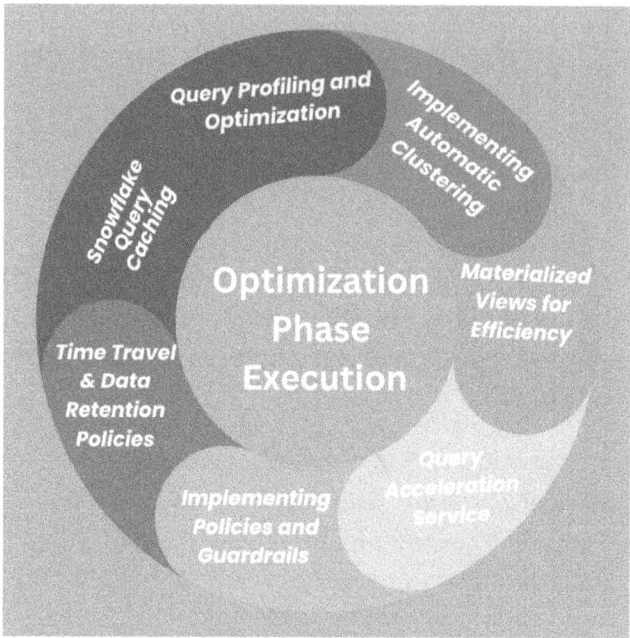

Figure 2-4. Executing the Optimization phase

Query Profiling and Optimization

Effective query profiling and optimization are critical for maximizing Snowflake performance while minimizing costs. By identifying inefficiencies in query execution and resource allocation, organizations can enhance both computational efficiency and financial accountability.

- **Leverage the Snowflake Query Profiler:** Use built-in profiling tools to analyze query performance and detect bottlenecks or inefficient patterns in workloads.

- **Optimize the query structure:** Adopt best practices such as limiting the use of SELECT *, filtering data early, and avoiding unnecessary computations to improve query efficiency.

- **Implement caching strategies:** Take advantage of Snowflake's result caching and materialized views to reduce redundant computations and improve execution time.

- **Monitor and refine usage patterns:** Regularly evaluate query logs to understand usage trends and tailor optimizations to frequently executed queries.

Query profiling and optimization not only reduce operational costs but also ensure that resources are used effectively, aligning Snowflake performance with organizational objectives.

Implementing Automatic Clustering

Automatic clustering in Snowflake is a helpful tool that organizes data to make queries faster and less expensive. Instead of manually setting up and maintaining clustering, Snowflake watches how data is used and automatically adjusts its organization to improve performance.

Here is why automatic clustering is beneficial:

- **Faster queries:** When tables are clustered, Snowflake uses metadata to find only the parts of the table that are needed for a query. This avoids scanning the whole table, saving time and reducing costs.

- **Adaptable data organization:** As data changes or grows, Snowflake automatically updates the organization of the table to match new patterns, keeping performance steady without extra work for administrators.

By using automatic clustering, businesses can save effort while keeping their important tables optimized for both speed and cost.

Using Materialized Views for Efficiency

Materialized views in Snowflake provide a powerful way to optimize performance and minimize compute costs, particularly for queries that are frequently executed or computationally demanding. Unlike standard views, which reprocess the underlying query each time they are accessed, materialized views store precomputed results. This enables Snowflake to deliver data instantly, avoiding repetitive computations and significantly improving query efficiency.

The key benefits of materialized views include:

- **Accelerated data access:** Queries involving complex joins, aggregations, or large datasets benefit greatly from materialized views. By using precomputed results, Snowflake reduces query execution times and enhances responsiveness for users.

- **Optimized resource utilization:** Organizations can target high-frequency, compute-intensive queries and transform them into materialized views. This approach lightens the load on virtual warehouses, leading to more efficient resource use and noticeable cost reductions.

- **Automated maintenance:** Snowflake ensures that materialized views stay current by automatically refreshing them whenever the underlying data is updated. This hands-free process guarantees the accuracy and reliability of query results without requiring manual oversight.

Materialized views are especially valuable for scenarios such as reporting, dashboarding, and analytics, where similar queries are repeatedly executed. When employed strategically, they deliver substantial improvements in both performance and cost-effectiveness within the Snowflake ecosystem.

Enhancing Performance with Query Acceleration Service (QAS)

Snowflake Query Acceleration Service (QAS) offers a robust solution for optimizing the performance of critical, high-priority queries by dynamically scaling compute resources. Unlike traditional methods that require fixed warehouse sizing, QAS enables on-demand compute bursts, ensuring that time-sensitive workloads are executed swiftly without the need for permanent overprovisioning.

The key benefits of QAS include:

- **Prioritization of critical workloads:** QAS is particularly beneficial in environments with varied and unpredictable query demands. Organizations can prioritize essential tasks, such as executive dashboards, real-time analytics, and customer-facing applications, ensuring they are completed promptly without disrupting other operations.

- **Adaptive and efficient resource allocation:** Unlike maintaining large, always-on warehouses, QAS allocates extra compute power only when necessary. This dynamic elasticity allows organizations to meet peak query demands efficiently while avoiding the expenses of unused capacity.

- **Enhanced user experience and business agility:** Faster query execution enhances the experience of data users, including analysts, data scientists, and business professionals who rely on timely insights for decision-making. By minimizing delays and improving responsiveness, QAS fosters a more agile, data-driven organization.

By incorporating QAS into a Snowflake strategy, organizations can strike a balance between high performance and cost efficiency, ensuring that critical workloads receive the attention they require without compromising overall operational effectiveness.

Leverage Time Travel and Data Retention Policies

Time Travel and data retention features in Snowflake help balance data accessibility with cost efficiency. These tools enable access to historical data for audits, recovery, and analysis—without duplicating tables or relying on costly backups.

- **Access to historical data:** Time Travel allows retrieval of previous data versions without manual backups or redundant tables. This simplifies recovery and auditing while reducing storage needs.

- **Retention period management:** Defining retention policies ensures only relevant historical data is kept. Aligning retention with business and compliance goals helps minimize storage use and eliminate outdated data.

- **Cost-effective recovery and auditing:** Built-in Time Travel and fail-safe features support reliable data recovery and auditing without external tools. This reduces compliance costs and ensures that critical data remains available.

By applying these features strategically, organizations can maintain a lean, compliant, and cost-effective data environment that preserves essential historical insights.

Using Snowflake Query Caching

Query caching in Snowflake helps reduce compute costs and improve performance. By reusing previously computed results and metadata, it minimizes redundant processing and speeds up response times, especially for repetitive queries.

- **Reusing query results:** Snowflake caches results from recent queries. If the data hasn't changed, rerunning the same query returns the cached result instantly, reducing compute usage and latency.

- **Metadata caching:** Snowflake also caches metadata like table structures and execution plans. This speeds up query parsing and planning, lowering computational overhead.

- **Cost efficiency:** Caching is especially valuable for repeated queries in dashboards, reports, or ad hoc analysis. Avoiding recomputation leads to significant savings in credit usage while maintaining fast performance.

To maximize the benefits of caching, teams should:

- Encourage consistent query patterns.
- Avoid unnecessary variations in query structures.
- Monitor cache hit rates as part of their performance optimization strategy.

Snowflake query caching, when utilized effectively, can become a cornerstone of a cost-conscious and performance-driven data strategy.

Note More detailed optimization strategies are introduced beginning in Chapter 5, with a focus on specific areas such as storage, compute, cloud services, AI, and data availability.

Summary

This chapter introduced a structured approach to FinOps in Snowflake, emphasizing the integration of financial discipline into technical operations. By progressing through stages of insight, control, and efficiency, organizations can better manage cloud costs while aligning usage with business priorities. Rather than relying solely on tools, the focus is on promoting a culture of accountability, encouraging informed decision-making, and embedding cost awareness into everyday practices.

By integrating financial discipline into the way Snowflake is used and governed, organizations can ensure that their investments are not only controlled but also optimized to deliver meaningful business outcomes. The next chapter builds on this foundation by exploring how visibility and transparency serve as the starting point for effective cost management and informed decision-making.

CHAPTER 3

Visibility and Transparency

Overview

Visibility and transparency are the foundation of FinOps. Access to critical usage data is essential for making informed financial decisions. This chapter explores a data-driven approach to help application owners and senior management optimize costs and improve operations. By leveraging actionable insights and streamlined reporting, organizations can foster a culture of accountability and enhance business performance. Specifically, this chapter examines how usage visibility can significantly increase return on investment (ROI) when using Snowflake.

The essential visibility and transparency strategies covered in this chapter include:

> **Customized reporting:** Tailored reports for different business stakeholders to highlight critical usage data.
>
> **Pricing awareness:** Detailed insights into low-level pricing models and tools to help stakeholders understand their cost consumption.
>
> **Snowflake metadata views:** Polished views from Snowflake's usage schemas to provide clear usage details from multiple perspectives.

By the end of this chapter, you will be equipped to visualize resource usage comprehensively, account for cost consumption, and transparently define accountability.

CHAPTER 3   VISIBILITY AND TRANSPARENCY

Enhanced Reporting and Continuous Evolution

To meet ever-evolving data requirements, it is imperative to anticipate new processes and provide stakeholders with up-to-date insights. This level of agility is essential for driving impactful change and ensuring the success of your FinOps initiatives.

Content automation plays a pivotal role in scaling FinOps insights across the organization. By automating the delivery of accurate data to the appropriate stakeholders, you can streamline processes, reduce manual effort, and enhance accuracy. This, in turn, minimizes errors and ensures data consistency. As illustrated in Figure 3-1, Snowflake usage metrics offer comprehensive insights into the usage of each service. The effective utilization of these metrics is crucial for achieving cost efficiency, and this section will elaborate on the methodologies to do so.

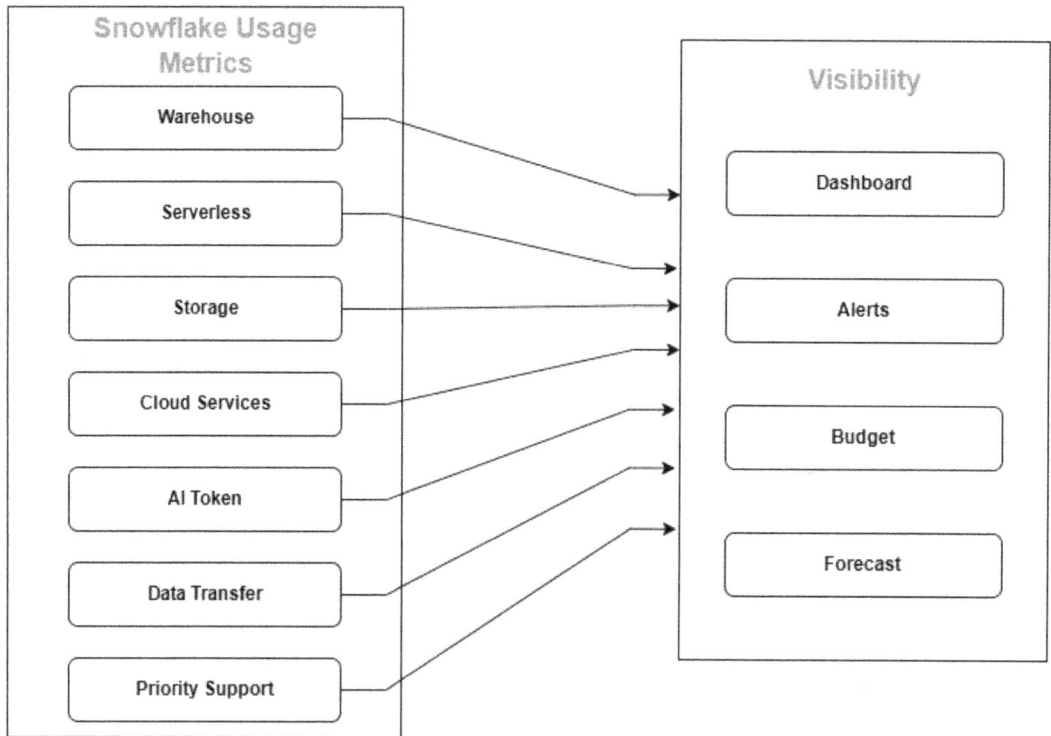

Figure 3-1. Snowflake metrics for monitoring

Scaling FinOps insights involves automating data delivery to enhance efficiency and foster a culture of ownership and efficiency in the organization. Here are some best practices for achieving efficient data visibility:

- **Shift information left:** Equip teams with the information and knowledge they need to make informed decisions independently.

- **Automate wherever possible:** Streamline processes through automation to ensure timely delivery of insights.

- **Be adaptable:** Stay agile and responsive to feedback and changes in data reporting, visualizations, and other transformations.

- **Promote understanding:** Turn data accessibility into usability for stakeholders, driving actionable insights.

While visibility best practices are set in place, assigning ownership to appropriate application owners leads to the next action.

Empowering Application Owners

Application owners benefit from having clarity on their usage and efficiency metrics on a regular basis. These reports empower them to integrate financial data into their decision-making, leading to better resource allocation. To facilitate this, consider setting up automated weekly and monthly reports that include the following key spending and savings-related information:

> **Application environment cost:** A breakdown of expenditures for each application environment, providing owners with insight into spending allocation.
>
> **Time-based trends:** Monthly cost variation data to aid in forecasting and budgeting.
>
> **Realized cost avoidance:** Metrics quantifying savings achieved through optimal operations and efficient resource consumption.
>
> **Efficiency scores:** Insights into resource utilization, enabling owners to identify KPIs for improvement.

Providing the latest usage metrics in the most accessible way helps application owners prioritize optimization efforts and maximize ROI from their business unit spending.

Senior Management and Wider Stakeholders

Financial funding is managed by senior management groups, including technology finance teams, procurement, executives, and senior directors. These groups participate in major organizational decisions. For this audience, we focus on key data points:

> **Year-to-date actual vs. budget:** A comprehensive overview of spending compared to budget expectations.
>
> **Variance analysis:** The difference between planned and the actual spend providing context of any overspend.
>
> **Commitment tracking:** Monitoring financial commitments to ensure alignment with strategic objectives.
>
> **Cost avoidance opportunities:** Identifying potential cost reductions through improved decision-making.
>
> **Actionable insights:** Clear recommendations for cost optimization and efficiency improvements.

By providing leadership with forecast metrics and vertical tracking, they can effectively allocate funding for future operations, measure usage against the business value that each vertical delivers, and identify areas of wasteful resource allocation across projects

The Snowflake Pricing Model

Controlling Snowflake costs requires understanding the "credits" system that's used for its usage-based pricing. Credit prices vary depending on the Snowflake edition and usage across different layers (compute, storage, and data transfer). Since each edition offers different features, understanding credit consumption helps manage expenditures.

Snowflake credits are the primary unit of measurement for resource consumption on the Snowflake platform, representing charges for using compute resources—Virtual Warehouse Compute, Serverless Compute, and Cloud Services Compute. Credit consumption depends on the virtual warehouse size; larger warehouses use more processing power and, therefore, consume more credits per hour.

Snowflake credits can be likened to fuel in a car. A more powerful engine (a larger warehouse) consumes more fuel (credits) per mile (hour of usage). Similar to a gas station billing based on gallons used, Snowflake uses a credit-to-dollar conversion rate to calculate the final bill.

The cost of each credit depends on four factors:

- Snowflake edition
- Cloud provider
- Hosting region
- On-demand/pre-purchase

Careful decisions must be made when selecting these factors, as they determine cost, balancing availability, security, and operational expenditure.

Collection of Performance and Usage Data

To maximize the benefits of the Snowflake platform, proactively monitor its activities and workloads. A solid understanding of current and historical usage data enables trend identification, resource optimization, and making informed decisions regarding total expenditure. Snowflake provides two schemas containing usage and cost data.

- ORGANIZATION_USAGE: Schema provides cost information for all accounts in the organization.
- ACCOUNT_USAGE: Schema provides similar information for a single account.

Data Latency

As a result of pulling the data from Snowflake's own internal metadata store, the usage views have an inherent delay:

– Account usage data has a latency of 45 minutes to three hours.

– Organization usage data has a latency of two hours to 72 hours.

Historical Data Retention Period

These views are retained for one year (365 days). To meet compliance requirements and analyze historical usage patterns and workload trends, it is important to retain usage data for longer than one year. It's important to list the most critical metrics that require historical data, which helps in developing trend analysis and forecasting predictions.

Build Historical Data Repository

As illustrated in Figure 3-2, you need to identify key catalog tables and store them in managed schemas (e.g., FOCUS_DATA and FOCUS_VIEW). Then you should engineer a scheduled job that captures usage data from these two database schemas and stores it according to your organization's retention needs (e.g., 3-5 years).

Metrics ranging from query history to warehouse current usage provide valuable insight in identifying recurring outliers, detecting resource waste, supporting audit and compliance requirements, and many other benefits.

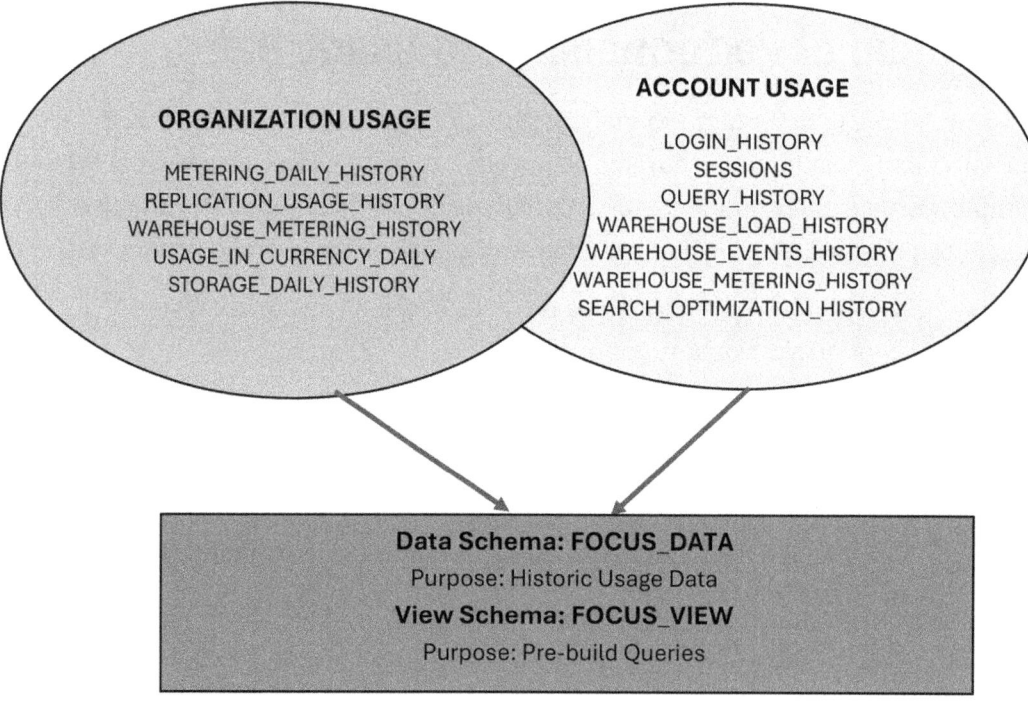

Figure 3-2. *Snowflake usage and cost historical data repository*

CHAPTER 3 VISIBILITY AND TRANSPARENCY

> **Note** Throughout this book, we use the ACCOUNT_USAGE and ORGANIZATION_USAGE views from the SNOWFLAKE database, considering general user usage.

You can also develop customized views to get the needed data from dictionary views.

FOCUS Prebuilt Views

To promote transparency in usage, it's essential to give users access to the data that matters most to them. One effective way to do this is to create customized, aggregated views, as shown in Table 3-1, that are tailored to specific business units or teams.

Table 3-1. FOCUS Prebuilt Views

View	Description
USER_QUERY_HISTORY	Performance metrics on user queries
USER_CREDIT_USAGE	Credit usage and associated cost for user query
USER_STORAGE_USAGE	Storage usage data for workspace schema
USER_AI_USAGE	AI and ML usage by user
USER_SERVERLESS_USAGE	Serverless features usage by user
USER_UNUSED_TABLES	Unused tables and relevant storage per owner
OBJECT_ACCESS_HISTORY	Object access history

These views can be designed to provide access to usage metrics for limited periods (e.g., the last 30 days), ensuring users have the insights they need to make informed decisions.

Snowflake Service Cost

This section provides an example and discusses the different components of Snowflake service costs to facilitate a better understanding of billing. The primary pricing factors affecting Snowflake's service costs are outlined in this section.

Snowflake's services have three primary pricing factors and one support model that contribute to the total cost of using the service:

- **Compute:** The biggest cost component in a Snowflake invoice
- **AI and ML:** Charged based on tokens used
- **Data storage:** On-demand storage cost
- **Data transfer:** Data movement cost (cross account/region)
- **Priority support:** Charged based on yearly spend on Snowflake services

To provide a clearer understanding, Figure 3-3 simulates a billing statement and the cost distribution of each service within the overall cost. Such a report can help trace where your dollar is spent within Snowflake.

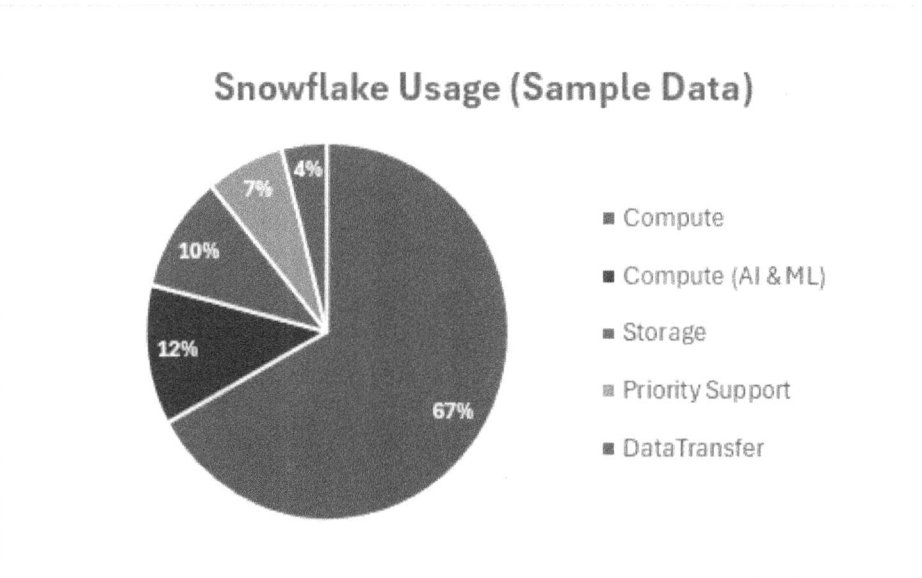

Figure 3-3. Snowflake usage report

Compute Service Cost

As shown in Figure 3-3, most costs are incurred in the compute layer, significantly outweighing costs in other layers. Therefore, optimizing this layer for performance and cost efficiency is crucial.

All compute usage is measured in Snowflake credits. The cost of using compute resources is calculated by multiplying the credit price by the number of credits consumed. Snowflake has three main types of compute resources that consume credits:

- Virtual warehouses
- Cloud services
- Serverless services

Virtual Warehouses

Virtual warehouses typically represent the most significant portion of Snowflake spending. These user-managed compute resources consume credits while processing data. Snowflake uses per-second billing (with a 60-second minimum per warehouse start), meaning that warehouses are billed only for the credits consumed during their active use based on their size.

Virtual warehouses features include:

> **Elasticity:** Rapidly scale up to handle large workloads and scale down as demand decreases.
>
> **Instant-on:** They start running in a short span when you need them.
>
> **Auto-suspend and auto-resume:** Automatically pause when idle and resume when a query is executed. This per-second metering ensures that you only pay for active usage.

Figure 3-4 illustrates the different warehouse sizes offered by Snowflake, along with the credit usage per hour. For Snowpark workloads, Snowflake provides a memory-optimized warehouse, known as the Snowpark Optimized Warehouse. As shown, there is an additional 1.5x premium for using these warehouses, which are designed to offer large memory capacities for machine learning workloads.

CHAPTER 3　VISIBILITY AND TRANSPARENCY

Warehouse Size (Credits / Hour)

Size	Standard	Snowpark,Optimized
XS	1	N/A
S	2	N/A
M	4	6
L	8	12
XL	16	24
2XL	32	48
3XL	54	96
4XL	128	192
5XL	256	384
6XL	512	786

Figure 3-4. Warehouse size (credits/hour)

Cloud Services

Cloud service costs cover a range of backend systems that support your Snowflake operations, including:

- Authentication
- Metadata management
- Access control
- Query compilation
- Executing commands (like SHOW commands)
- Caching results for faster access

Charges for the cloud services layer apply when daily cloud services resource consumption exceeds 10% of daily warehouse usage. This is known as *cloud services adjustment*. However, usage related to serverless features and Snowpark Container Services (SPCS) compute is excluded from this adjustment. The cloud services adjustment prevents overcharging for the backend systems supporting Snowflake operations by activating when their usage becomes significant relative to warehouse usage.

Now we explain how credit adjustment works with the example shown in Table 3-2.

Table 3-2. Cloud Services Credit Calculation

Date	Compute Cost	Cloud Services Cost	Credit Adjustment for cloud services	Cloud Services Cost Due	Total Cost
Jan 1	120	25	12	13	133
Jan 2	90	12	9	3	93
Jan 3	160	11	16	0	160
Jan 4	100	9	10	0	100

In the example provided, the compute cost on January 1st is 120, and the cloud services cost is 25. Ten percent of the compute cost is 12. Therefore, 12 is the adjustment amount, resulting in a final cost of 133(120 + (25 − 12)). Applying the same formula to January 2nd, with a compute cost of 90 and a cloud services cost of 12, results in a final cost of 93(90 + (12 − 9)).

Consider Oct 3rd and 4th, where the cloud services cost is less than 10% of the compute cost. Hence, the cloud services cost will be considered an adjustment, and it will be waived.

Serverless Services

Snowflake offers several serverless features, such as Snowflake data sharing and Snowflake data exchange. These features use Snowflake-managed computing resources, and billing is based on the number of compute hours used. The number of credits consumed per compute hour varies depending on the specific serverless feature.

One of the key benefits of Snowflake's serverless offerings is their "set it and forget it" nature. With this model, resource management and workload right-sizing become automatic. Snowflake manages resources, scaling them up or down based on demand for a small markup. In essence, you configure your serverless capabilities, and Snowflake handles resource scaling without requiring constant monitoring or configuration changes for an additional premium cost in credits, as shown in Table 3-3.

Table 3-3. Serverless Features Credits Per Compute Hour

Serverless Feature Table	Snowflake-Managed Compute	Cloud Services
Clustered Tables 2 1	2	1
Copy Files5 2 N/A	2	N/A
Data Quality Monitoring	2	1
Hybrid Tables Requests	1	1 In addition: 1 Credit per 30GB read 1 Credit per 7.5GB write
Logging	1.25	N/A; instead charged 0.28 Credits per 1000 file batches
Materialized Views maintenance	2	1
Organization Usage	1	1
Query Acceleration	1	N/A
Replication	2	0.35
Search Optimization Service	2	1
Serverless Alerts	1.2	1
Serverless Tasks	0.9	1
Snowpipe	1.25	N/A; instead charged 0.06 Credits per 1000 files
Snowpipe Streaming	1	N/A; instead charged at an hourly rate of 0.01 Credits per client instance

These serverless features enable you to design specific use cases for greater efficiency and optimized Snowflake costs.

Snowflake AI and ML Costs

Snowflake offers AI features, known as Snowflake AI features. These features run on Snowflake-managed computing resources, and customers use Snowflake credits to pay for them. Large language model (LLM) functions are charged based on the number of tokens processed.

- **Snowflake Cortex** is a suite of AI capabilities built on LLMs designed to understand unstructured data, respond to freeform inquiries, and deliver intelligent, concierge-like experiences.

- **Snowflake ML** allows users to build their own machine-learning models.

A *token* is the smallest unit of text that an AI model processes. The conversion of raw input or output text into tokens depends on the specific Cortex model. Billing is based on the number of tokens processed, which, depending on the model, may include only input tokens or both input and output tokens.

CHAPTER 3   VISIBILITY AND TRANSPARENCY

A token can be understood as a "unit of text" that the model processes. For further understanding, see Figure 3-5.

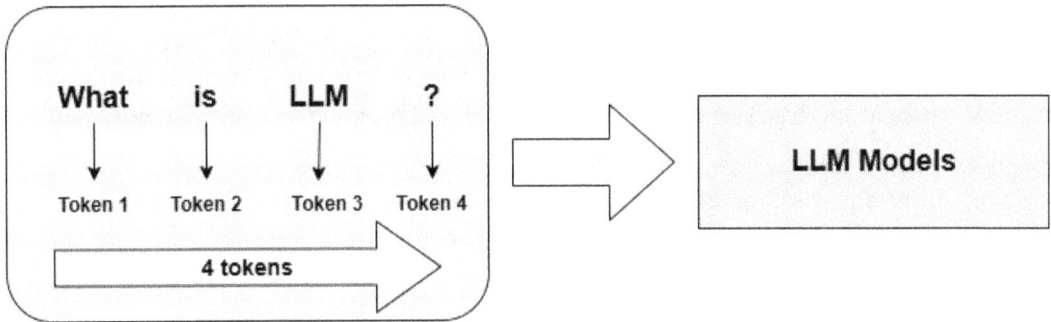

Figure 3-5. *Snowflake AI/ML tokens*

Managing Cortex LLM Costs

As we write in this book, Snowflake recommends using a warehouse no larger than Medium when calling Snowflake Cortex LLM functions. Using a larger warehouse than necessary does not improve performance, but can lead to unnecessary costs and an increased risk of throttling.

To track credits used for AI services, including LLM functions in your account, use the METERING_DAILY_HISTORY view.

Storage Service Cost

Storage costs are priced by the terabyte per month and encompass all customer data held in a Snowflake account, typically representing a small portion of the total Snowflake bill.

Storage cost represents the cost of:

- Permanent tables include historical data for Time Travel
- Fail-safe used for recovery
- Files staged for data loading/unloading (stored compressed or uncompressed)
- Clones of tables that reference data deleted in the table that owns the clones

CHAPTER 3 VISIBILITY AND TRANSPARENCY

Data Storage Billing Calculation

Storage pricing is based on the average terabytes of customer data stored in the Snowflake account per month. This average is calculated by taking hourly snapshots of all customer data and averaging them across each day. This daily average is displayed in the Snowflake service. The monthly charge is based on the average calculated across the number of calendar days in the month. Snowflake charges are based on compressed storage; Snowflake measures the size of your data after compression, not its original size.

Formula:
```
For a given month = (((SUM(Daily Average Storage Used) of whole month /
                    number of days in the month)) * Storage Rate)
```
Example

Assume the Snowflake account is at capacity. It's an AWS–US West snowflake account and the on-demand price is $23/TB/month.

Daily average storage used is as follows:

> Day 1: 50 TB
>
> Day 2: 60 TB
>
> Day 3: 100 TB

No new data was loaded in the remaining days; therefore, the usage between days 3 and 30 is 100 TB. Using the calculation in Table 3-4, you can derive the total cost for storing the data for a month.

Table 3-4. Storage Billing Calculation

Storage Billing Calculation	Storage(TB)	No. of Days	Avg. Storage Used (TB)
Day1	50	1	50
Day2	60	2	120
Day3	100	28	2800
SUM(Daily Average Storage Used)			2970
Number of days in the month		30	
Storage Per day			99.00
Storage Rate/TB/month			$23.00
Total Cost			$2,277.00

Data Transfer (Ingress or Egress) Service Cost

Data transfer involves moving data into (ingress) and out of (egress) Snowflake.

Snowflake's data transfer pricing is based on rates set by the cloud provider and includes a per-byte free for data egress when users transfer data from their Snowflake account to a different region within the same cloud platform or to a different platform.

Here is how data transfer costs work in Snowflake:

- Snowflake does not charge for data loads (ingress).

- Transferring data between the same region in the cloud provider is free.

- Only specific Snowflake features incur data transfer costs (unloading data, replicating data, using external functions, cross-cloud auto-fulfillment, etc.).

- Data egress charges do not apply when a Snowflake client or driver retrieves query results, even if those happen across cloud platforms or regions.

Setting up a secondary site for your business continuity plan (BCP) incurs compute and data transfer costs. Planning the data volume at the secondary site is crucial, as it directly impacts daily data transfer costs. You can use the following query to measure the credit spent on the data replication setup:

```
SELECT
      start_time::date as date
      ,sum(credits_used) as credits_used
FROM snowflake.account_usage.replication_usage_history
WHERE start_time >= dateadd(year,-1,current_timestamp())
GROUP BY 1
ORDER BY 2 DESC;
```

Note The replication usage cost is reflected for customers who have set up a replication secondary site and enabled data transfer.

Priority Support Cost

Snowflake's priority support usually varies from the usual platform charges and is based on annual spending on Snowflake services. The current pricing structure is broadly based on this model:

> **Base support**: Bundled with all Snowflake accounts, this service allows you to engage in support for critical issues but with a standard response time.
>
> **Priority support**: This is charged as a percentage of annual Snowflake spending, with the specific rate depending on account size, the chosen support tier, and any negotiations with Snowflake.

Focused tiers typically cost more but include a dedicated technical account manager (TAM), faster response times, responsive monitoring, and enhanced troubleshooting.

For example, priority support can range from 5% to 10% of the total annual Snowflake spend, depending on service levels and location. You can use the following query to determine the total cost spent for priority support by your organization.

```
SELECT
  DATE_TRUNC('MONTH', USAGE_DATE::DATE) AS "Usage Month",
  SUM(USAGE) AS "Total Usage",
  SUM(USAGE_IN_CURRENCY) AS "Total Cost"
FROM
  SNOWFLAKE.ORGANIZATION_USAGE.USAGE_IN_CURRENCY_DAILY
WHERE USAGE_DATE >= DATE_TRUNC('MONTH', DATEADD(month,-3,CURRENT_TIMESTAMP()))
  AND BILLING_TYPE = 'priority support'
GROUP BY ALL
ORDER BY 1;
```

> **Note** For the most accurate and up-to-date pricing information for the Snowflake paid support model, contact Snowflake directly, as pricing varies based on usage patterns, geography, and contractual agreements. Pricing for these support packages typically falls under a custom agreement or an enterprise-level service contract.

CHAPTER 3   VISIBILITY AND TRANSPARENCY

Exploring Usage Costs

UI-Based Cost Management

The cost management interface provides a comprehensive toolkit and insights to manage your cash flow. It gives you complete visibility at the organization and account level and you can get the insights to make timely decisions and optimize cost proactively.

Organization Level Usage

The Organization overview gives you a broad look at your spending across multiple accounts:

- **Spend summary:** Instant view of essential spend metrics, including total spend in USD, remaining balance, average monthly spend, and total accounts.

- **Contract overview:** Dive into detailed insights about your contract usage and forecasts. You'll see a comparison of actual spend vs. forecasted spend, track contract progress, and understand contract duration.

- **Account spend summary:** View a breakdown of your top accounts by spend, complete with details like accumulated spend for the period, current month's spend, last month's spend, and your monthly average.

This experience gives you a clear, unified view of your financial landscape, helping you manage and optimize spend more effectively

Account Level Usage

Account level provides an instant visual dashboard from which you can have detailed views on your financial activity:

- **Monitor account spend:** Keep track of your spending with an easy-to-understand overview, comparing actual costs with your set budgets to stay on top of your financial goals.

- **Forecast spend:** Anticipate future expenses based on your budget allocations, helping you ensure that your financial plans stay on track and within limits.

- **Identify top areas by spend:** Highlight the areas where you're spending the most, such as on warehouses, queries, or databases by storage, so you can focus on what matters most.

- **Optimize spend:** With the upcoming Cost Insights feature, you'll be able to uncover opportunities for cost savings, ensuring that every dollar is used as effectively as possible.

As shown in Figure 3-6, through cost management, you can determine the total spend for a period and the top warehouses that contributed to this spend.

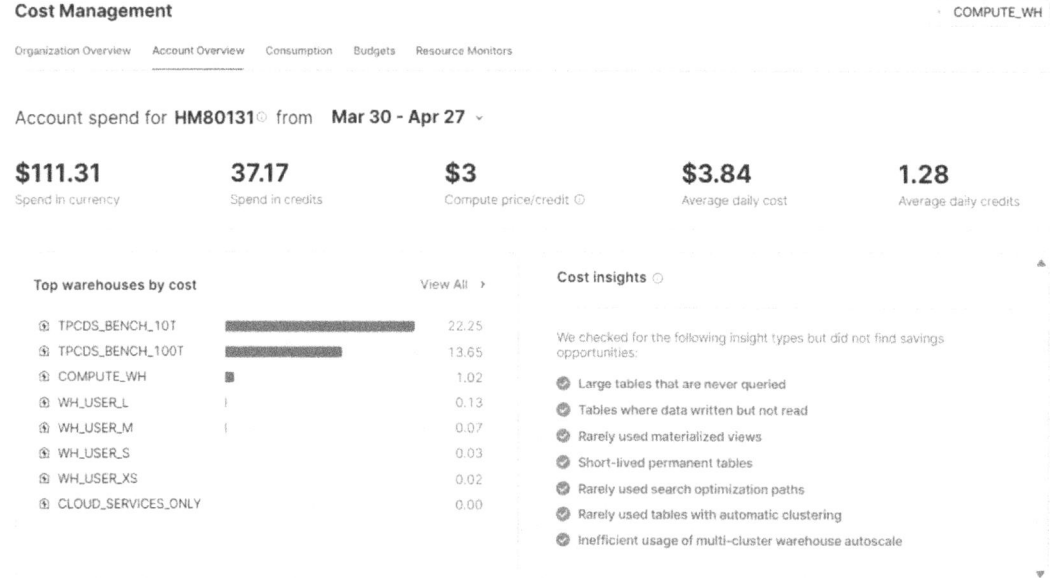

Figure 3-6. Cost management

Cost Insights (Recommendation)

Cost management now provides additional insight for saving opportunities.

- Large tables that are never queried
- Tables where data is written but not read
- Rarely used materialized views

CHAPTER 3 VISIBILITY AND TRANSPARENCY

- Short-lived permanent tables
- Rarely used search optimization paths
- Rarely used tables with automatic clustering
- Inefficient use of multi-cluster warehouse auto-scale

Figure 3-7. Most expensive queries

As shown in Figure 3-7, the analysis shows the most expensive queries with the total number of executions, average execution time, and warehouse name, along with the user who executed the query.

Warehouse Usage

Cost management shows a summary of warehouse consumption for the given period, as shown in Figure 3-8.

Figure 3-8. Cost management – credit consumption

Figure 3-8 shows the warehouse name and the credits consumed by that warehouse. It also provides insights into the consumption patterns of each warehouse.

CHAPTER 3 VISIBILITY AND TRANSPARENCY

Drilling Down into Incurred Costs

You can use the Consumption page to drill down to view the credits used by the budget of Snowflake for any given day, week, or month, as shown in Figure 3-9.

Figure 3-9. Cost management – consumption insights

The insights display the credits consumed over a specific period. You can use filters to derive consumption based on various criteria, such as period, user, specific resource, and service type.

Account-Level Budgets

You can also set up an account-level budget to limit overspending and get projected spending for upcoming days.

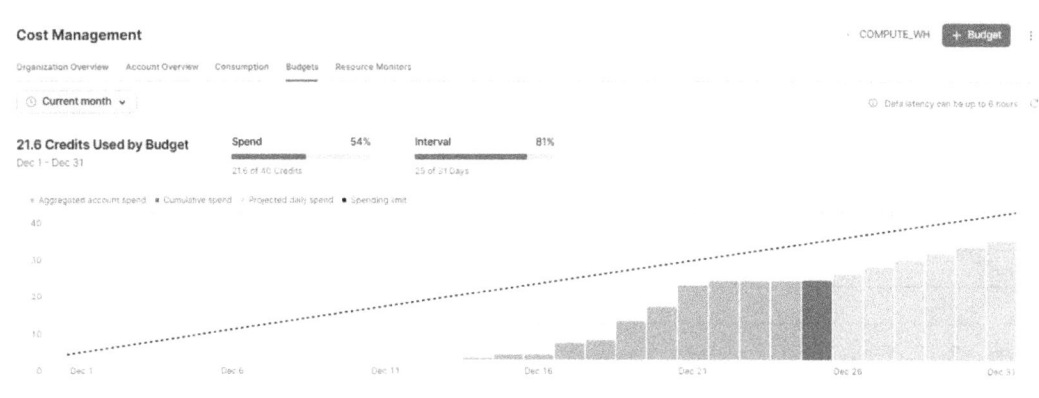

Figure 3-10. Cost management - account level budgeting

CHAPTER 3 VISIBILITY AND TRANSPARENCY

As you can see in Figure 3-10, the cost management has a section for budget, which shows how much credit has been spent at a particular time. Having some credits allotted as the threshold limit, this budget column shows how many credits have been consumed so far.

Using Predefined Queries

By running queries against the ACCOUNT_USAGE and ORGANIZATION_USAGE views, you can dive deep into your cost data and create custom reports and dashboards tailored to your needs. To do so, you use this query (the following image shows an example):

```
--Daily credit usage and usage in currency
SELECT account_name,billing_type,service_type,
  ROUND(SUM(usage_in_currency), 2) as usage_in_currency
FROM focus_data.organization_usage.usage_in_currency_daily
WHERE usage_date > DATEADD(month,-1,CURRENT_TIMESTAMP())
GROUP BY ALL
ORDER BY 4 desc;
```

	ACCOUNT_NAME	BILLING_TYPE	SERVICE_TYPE	USAGE_IN_CURRENCY
1	PG56314	CONSUMPTION	WAREHOUSE_METERING	151.24
2	PG56314	CONSUMPTION	STORAGE	12.02
3	PG56314	CONSUMPTION	CLOUD_SERVICES	11.00

Compute Service Cost

Virtual Warehouses

Snowflake virtual warehouse costs are calculated based on the size of the warehouse and its active time. Charges are incurred when the warehouse is running, with pricing varying by the warehouse size (e.g., X-Small, Small, Medium).

Use this query to get the virtual warehouse usage cost (the following image shows an example):

```
-- Compute Service Usage by day
SELECT
  start_time::date as the_dt,
  warehouse_name,
  SUM(credits_used_compute) AS credits_used_compute_sum
FROM focus_data.account_usage.warehouse_metering_history
WHERE start_time >= DATEADD(day, -7, CURRENT_TIMESTAMP())
  AND warehouse_id > 0  -- Skip "CLOUD_SERVICES_ONLY"
GROUP BY ALL
ORDER BY 1;
```

	THE_DT	WAREHOUSE_NAME	CREDITS_USED_COMPUTE_SUM
1	2024-11-15	COMPUTE_WH	149.17
2	2024-11-16	WH_USER_L	139.36
3	2024-11-16	WH_USER_XS	121.26
4	2024-11-16	WH_USER_S	124.04

Cloud Services Cost

Snowflake cloud services costs are primarily driven by resource consumption for tasks such as query processing, metadata management, and data operation orchestration. To better understand spending patterns, Snowflake's usage data can be queried to calculate the average hourly credit consumption across all warehouses.

Use this query to get the cloud services cost (the following image shows an example):

```
SELECT
    WAREHOUSE_NAME
    ,SUM(CREDITS_USED) as CREDITS_USED
    ,SUM(CREDITS_USED_CLOUD_SERVICES) as CREDITS_USED_CLOUD_SERVICES
    ,(SUM(CREDITS_USED_CLOUD_SERVICES)/SUM(CREDITS_USED))*100 as PERCENT_CLOUD_SERVICES
FROM FOCUS_DATA.ACCOUNT_USAGE.WAREHOUSE_METERING_HISTORY
WHERE START_TIME >= DATEADD(month,-1,CURRENT_TIMESTAMP())
```

```
        AND CREDITS_USED_CLOUD_SERVICES > 0
GROUP BY 1
ORDER BY 4 DESC;
```

	WAREHOUSE_NAME	CREDITS_USED	CREDITS_USED_CLOUD_SERVICES	PERCENT_CLOUD_SERVICES
1	CLOUD_SERVICES_ONLY	115.00	4.00	3.48
2	WH_USER_M	107.09	5.00	4.67
3	WH_USER_L	124.36	5.00	4.02
4	WH_USER_XS	124.26	5.00	4.02
5	COMPUTE_WH	156.55	4.01	2.56

Serverless Service Cost

Snowflake's serverless compute feature allows users to run queries and manage workloads without the need to provision dedicated compute resources. This model is more flexible and scalable, given that it scales compute automatically with workload requests, but it does reflect a different cost structure. The cost is calculated by the second, rounded up to the nearest whole second, with the price varying depending on the spending behavior.

You'll need to analyze the costs based on the different compute sizes chosen during the execution. The details can be gathered through views, which are shown in Table 3-5.

Table 3-5. *Serverless Usage Views*

View	Description
MATERIALIZED_VIEW_REFRESH_ HISTORY	Credits consumed by refreshing materialized views
SEARCH_OPTIMIZATION_HISTORY	Credits consumed by the search optimization service
PIPE_USAGE_HISTORY	Credits consumed by Snowpipe
QUERY_ACCELERATION_HISTORY	Credits consumed by the query acceleration service
REPLICATION_USAGE_HISTORY	Database replication credit usage and number of bytes transferred
AUTOMATIC_CLUSTERING_HISTORY	Credits consumed by automatic clustering
CORTEX_FUNCTIONS_ USAGE_HISTORY	Credits consumed to call Cortex LLM functions

Use this query to get the serverless services usage cost.

```sql
--Serverless service credit usage
SELECT USAGE_DATE, SERVICE_TYPE,
       SUM(CREDITS_BILLED) AS CREDITS_BILLED
FROM FOCUS_DATA.ACCOUNT_USAGE.METERING_DAILY_HISTORY
WHERE SERVICE_TYPE IN (
              'AUTO_CLUSTERING',
              'DATA_QUALITY_MONITORING',
              'MATERIALIZED_VIEW',
              'PIPE',
              'QUERY_ACCELERATION',
              'REPLICATION',
              'SEARCH_OPTIMIZATION',
              'SERVERLESS_ALERTS',
              'SERVERLESS_TASK',
              'SNOWPIPE_STREAMING',
              'HYBRID_TABLE_REQUESTS')
GROUP BY ALL;
```

	USAGE_DATE	SERVICE_TYPE	CREDITS_BILLED
1	2024-11-16	SERVERLESS_TASK	6.00

AI Services Cost Usage

To track credits used for AI services, including LLM functions in your account, use this query on METERING_DAILY_HISTORY to get AI services usage cost (the following image shows an example):

```sql
SELECT USAGE_DATE, SERVICE_TYPE,
       ROUND(SUM(CREDITS_BILLED),2) AS CREDITS_BILLED
  FROM FOCUS_DATA.ACCOUNT_USAGE.METERING_DAILY_HISTORY
  WHERE SERVICE_TYPE='AI_SERVICES'
GROUP BY ALL
ORDER BY USAGE_DATE;
```

CHAPTER 3 VISIBILITY AND TRANSPARENCY

USAGE_DATE	SERVICE_TYPE	CREDITS_BILLED
2024-12-15	AI_SERVICES	4.00

Track Credit Consumption for LLM Functions

To view the credit and token consumption for each LLM function call, use this query on the CORTEX_FUNCTIONS_USAGE_HISTORY view (the following image shows an example):

```
SELECT *
  FROM FOCUS_DATA.ACCOUNT_USAGE.CORTEX_FUNCTIONS_USAGE_HISTORY;
```

START_TIME	END_TIME	FUNCTION_NAME	MODEL_NAME	WAREHOUSE_ID	TOKEN_CREDITS	TOKENS
2024-12-14 16:00:00.000 -0800	2024-12-14 17:00:00.000 -0800	COMPLETE	llama2-70b-chat	4	0.000014400	32
2024-12-14 16:00:00.000 -0800	2024-12-14 17:00:00.000 -0800	EMBED_TEXT	snowflake-arctic-e	4	0.000000960	32
2024-12-14 16:00:00.000 -0800	2024-12-14 17:00:00.000 -0800	CLASSIFY_TEXT	null	4	0.001248220	898
2024-12-14 16:00:00.000 -0800	2024-12-14 17:00:00.000 -0800	COMPLETE	mistral-7b	4	0.000018360	153
2024-12-14 16:00:00.000 -0800	2024-12-14 17:00:00.000 -0800	COMPLETE	snowflake-arctic	4	0.000349440	416
2024-12-14 16:00:00.000 -0800	2024-12-14 17:00:00.000 -0800	TRANSLATE	null	4	0.002178000	1452

Storage Service Cost

Snowflake storage charges are typically calculated on a per-terabyte and per-month basis, and Snowflake automatically compresses data to minimize storage space. Prices vary by region and cloud provider (AWS, Azure, or GCP). Total storage cost is the sum of costs associated with:

- Database table storage
- Fail-safe and Time Travel storage
- Hybrid table storage
- Staged file storage

Total storage cost is tracked using this query on the STORAGE_USAGE view (the following image shows an example):

```
SELECT USAGE_DATE,
       SUM ((((STORAGE_BYTES +
             STAGE_BYTES +
             FAILSAFE_BYTES +
```

```
                HYBRID_TABLE_STORAGE_BYTES
     )/1024)/1024)/1024)/1024 AS STORAGE_USAGE_IN_TB
FROM    FOCUS_DATA.ACCOUNT_USAGE.STORAGE_USAGE
GROUP   BY ALL;
```

USAGE_DATE	STORAGE_USAGE_IN_TB
1 2024-11-12	1.00
2 2024-11-13	1.00
3 2024-11-14	3.00
4 2024-11-15	3.01

Data Transfer Cost

Data transfer involves moving data into (ingress) and out of (egress) Snowflake.

Snowflake's data transfer pricing is based on rates set by the cloud provider and includes a per-byte free for data egress, when users transfer data from their Snowflake account to a different region within the same cloud platform or to a different platform.

Here is how data transfer costs work in Snowflake:

- Snowflake does not charge for data load (ingress).

- Transferring data between the same region and the cloud provider is free.

- Only specific Snowflake features incur data transfer costs (unloading data, replicating data, using external functions, cross-cloud auto-fulfillment, etc.).

- Data egress charges do not apply when a Snowflake client or driver retrieves query results, even if those happen across cloud platforms or regions.

Replication data transfer cost is tracked using this query on the REPLICATION_USAGE_HISTORY view:

```
SELECT
        start_time::date as date
        ,sum(credits_used) as credits_used
```

```
FROM focus_data.account_usage.replication_usage_history
WHERE start_time >= dateadd(year,-1,current_timestamp())
GROUP BY 1
ORDER BY 2 DESC;
```

Note The replication usage cost is reflected for customers who have set up a replication secondary site and enabled data transfer.

Priority Support Cost

Snowflake's priority support usually varies from the usual platform charges and is based on annual spending on Snowflake services. The current pricing structure is broadly based on this model:

> **Base support:** Bundled with all Snowflake accounts, this service allows you to engage in support for critical issues but with a standard response time.
>
> **Priority support:** This is charged as a percentage of annual Snowflake spending, with the specific rate depending on account size, the chosen support tier, and any negotiations with Snowflake.

Focused tiers typically cost more but include a dedicated technical account manager (TAM), faster response times, responsive monitoring, and enhanced troubleshooting.

For example, priority support can range from 5% to 10% of the total annual Snowflake spend, depending on service levels and location.

Use the this query to find the priority support cost for your organization:

```
SELECT
  DATE_TRUNC('MONTH', USAGE_DATE::DATE) AS "Usage Month",
  SUM(USAGE) AS "Total Usage",
  SUM(USAGE_IN_CURRENCY) AS "Total Cost"
FROM
  FOCUS_DATA.ORGANIZATION_USAGE.USAGE_IN_CURRENCY_DAILY
WHERE USAGE_DATE >= DATE_TRUNC('MONTH', DATEADD(month,-3,CURRENT_
TIMESTAMP()))
```

```
  AND BILLING_TYPE = 'priority support'
GROUP BY ALL
ORDER BY 1;
```

Note For the most accurate and up-to-date pricing information for the Snowflake paid support model, contact Snowflake directly, as pricing varies based on usage patterns, geography, and contractual agreement. Pricing for these support packages typically falls under a custom agreement or an enterprise-level service contract.

Tagging and Cost Attribution
Query Tag

Query tags are optional session-level parameters used to tag individual queries or sets of queries. Use QUERY_HISTORY to map queries to their corresponding tag values for various use cases. This provides insights into your query workload for monitoring, tuning, and optimization.

> **Efficient query tracking:** Monitor overall performance by tags, such as the aggregate number of query executions and total query duration.
>
> **Enhanced reporting:** Tags provide a systematic way to extract detailed reports on query performance and usage patterns.

The tag can be set at a session level (Account->User->Session) and can also be configured when issuing Snowflake queries from tools such as dbt, airflow, and so on.

Structure Your QUERY_TAG in JSON:

```
{
  "app": "airflow_daily_load",
  "environment": "production",
  "user": "airflow_user",
  "job": "sales_tran",
  "version": "v1.2"
}
```

CHAPTER 3   VISIBILITY AND TRANSPARENCY

Set query tags at the session level:

```
ALTER SESSION SET QUERY_TAG =  <<json query tag>>;
```

Set default tags for a user:

```
ALTER USER airflow_user SET QUERY_TAG = <<json query tag>>;
```

Use query tags for cost and performance monitoring:

```sql
SELECT
    try_parse_json(query_tag)['app']::string as app_name,
    try_parse_json(query_tag)['job']::string as job_name,
    count(*) as job_runs,
    avg(total_elapsed_time/60000) as avg_total_elapsed_min,
    avg(bytes_scanned/1024/1024/1024) as avg_bytes_scanned_gb
FROM focus_data.account_usage.query_history
WHERE try_parse_json(query_tag)['app']::string = 'airflow_daily_load'
AND try_parse_json(query_tag)['job']::string = 'sales_tran'
AND start_time >= date_trunc('month', dateadd(month,-3,current_
timestamp()))
AND warehouse_size is not null
GROUP BY ALL;
```

APP_NAME	JOB_NAME	JOB_RUNS	AVG_TOTAL_ELAPSED_MIN	AVG_BYTES_SCANNED_GB
airflow_daily_load	sales_tran	2	1.95	4.55

From Snowsight, apply a filter on the query tag to see the query attributes (see the following image).

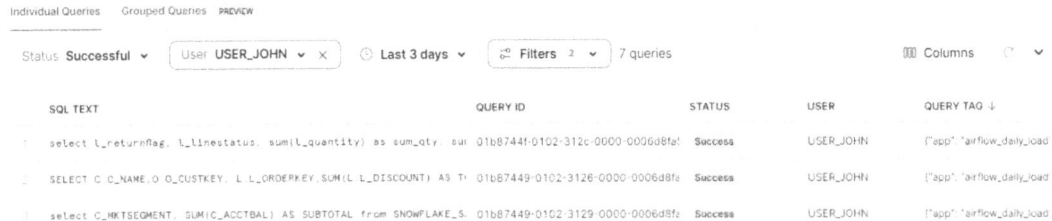

Object Tags

Object tags are used to tag persistent account objects such as users, roles, tables, views, and functions. Both query and object tags serve the common purpose of enabling more structured monitoring and increased visibility within your Snowflake account

```
ALTER WAREHOUSE WH_USER_XS SET TAG cost_center = 'finance';
ALTER WAREHOUSE WH_USER_S SET TAG cost_center = 'marketing';
```

Use the object tag to attribute the credit usage for queries running on the warehouses tagged to a specific cost center.

```
SELECT
    TAG_VALUE AS COST_CENTER,
    SUM(NVL(CREDITS_USED, 0)) AS CREDITS
FROM
    FOCUS_DATA.ACCOUNT_USAGE.WAREHOUSE_METERING_HISTORY
JOIN
    SNOWFLAKE.ACCOUNT_USAGE.TAG_REFERENCES
    ON WAREHOUSE_NAME=OBJECT_NAME
WHERE START_TIME>=DATEADD('DAYS',-7,CURRENT_DATE())
AND TAG_VALUE IN ('finance','marketing')
GROUP BY ALL;
```

	COST_CENTER	CREDITS
1	marketing	134.04
2	finance	129.33

Relative Cost Per Query

Organizations can associate Snowflake expenses with logical units (e.g. users, departments, and projects) to understand the usage costs. This benefits accounting and helps identify high-spending units, allowing for targeted implementation of controls and potential cost optimization.

CHAPTER 3 VISIBILITY AND TRANSPARENCY

Boost transparency: Understand the contribution of individual users or business units to overall costs.

Optimize chargebacks: Establish accurate chargeback models to ensure precise cost allocation to the appropriate teams.

The cost attributes per query are available in the QUERY_ATTRIBUTION_HISTORY view. However, this charge does not cover all credits incurred by query execution. For example, the following costs are not included in the query cost:

- Data transfer costs
- Storage costs
- Cloud services costs
- Costs for serverless features
- Costs for tokens processed by AI services

To estimate the cost of running a query, consider the following factors:

Size of the warehouse: Larger warehouses typically incur higher costs.

Number of clusters: More clusters can affect overall compute usage.

Concurrent queries: The number of concurrently running queries can impact resource allocation.

Auto-suspend rules: The warehouse's auto-suspend behavior influences costs.

Warehouse level cache usage: Effective cache utilization can reduce compute needs and costs.

Users can be mapped to business units/cost centers in a custom view, This mapping, combined with user-based costs, enables rolling up spending across needs and costs.

Use this query to determine the query attribute cost for a user (the following image shows an example):

```
--Query attribute cost for given user
SELECT query_id,user_name,start_time,warehouse_name,
    COALESCE(NULLIF(query_tag, ''), 'untagged') AS tag,
```

```
    credits_attributed_compute AS credits
  FROM focus_data.account_usage.query_attribution_history
  WHERE start_time >= DATE_TRUNC('MONTH', DATEADD(MONTH, -1,
CURRENT_DATE))
    AND user_name = 'USER_JOHN'
;
```

QUERY_ID	USER_NAME	START_TIME	WAREHOUSE_NAME	TAG	CREDITS
1	01b87412-0102-3126-0000-0006d8fad71!	USER_JOHN	2024-11-18 04:34:28.308 -0800	WH_USER_M	{ "app": "airflow_daily_load",	0.01628555556
2	01b87449-0102-3126-0000-0006d8fad78(	USER_JOHN	2024-11-18 05:29:46.661 -0800	WH_USER_S	{"app": "airflow_daily_load"}	0.07819137515
3	01b87449-0102-3129-0000-0006d8fa4e2(	USER_JOHN	2024-11-18 05:29:33.227 -0800	WH_USER_S	{"app": "airflow_daily_load"}	8.364336988e-05
4	01b87415-0102-3127-0000-0006d8fa2d2!	USER_JOHN	2024-11-18 04:37:09.008 -0800	WH_USER_XS	{ "app": "airflow_daily_load",	0.05362217593

Exploring the Billing Usage Statement

How to Locate the Usage Statement

Snowflake provides a monthly usage statement for customers with at least one active contract. This statement includes detailed usage information in currency, a summary of lifetime usage, and a breakdown of monthly usage in terms of charged credits and associated costs.

The monthly usage statement can be found and downloaded in Snowsight. Only users with ACCOUNTADMIN and ORGADMIN roles are authorized to view and download this statement.

Reconcile Usage Statement

You can reconcile each service type of usage in credits and currency value using this query (the following image shows an example):

```
SELECT
      DATE_TRUNC(month, usage_date) AS usage_month,
      CONCAT(account_locator,'-',region) AS account_name,
      usage_type AS usage_category,
      SUM(usage) AS units_consumed,
      SUM(usage_in_currency) AS total_usage
```

```
FROM focus_data.organization_usage.usage_in_currency_daily
WHERE TRUE
    AND usage_month >= DATEADD(year,-1,CURRENT_TIMESTAMP())
GROUP BY ALL
ORDER BY 1,2,3;
```

USAGE_MONTH	ACCOUNT_NAME	USAGE_CATEGORY	UNITS_CONSUMED	TOTAL_USAGE
2024-11-01	HA38039-AWS_US_EAST_2	adj for incl cloud services	-0.019	-0.06
2024-11-01	HA38039-AWS_US_EAST_2	cloud services	0.019	0.06
2024-11-01	HA38039-AWS_US_EAST_2	compute	11.672	35.01
2024-11-01	HA38039-AWS_US_EAST_2	storage	0.002	0.04

Summary

This chapter has covered Snowflake service costs and explained how to monitor and explore them using Snowsight and prebuilt queries. Snowsight provides a graphical interface for viewing and analyzing cost data, while prebuilt queries allow for detailed breakdowns. We also discussed how tagging and cost attributes can decompose costs by resource, role, or department, enabling users to identify key cost drivers and manage usage accordingly. These tools significantly improve the visibility into the contribution of various workloads to total costs. The next chapter covers cost control capabilities and explains how to design budgets in Snowflake for better cost management.

CHAPTER 4

Taking Control of Your Snowflake Spend

Overview

The Control phase in FinOps focuses on establishing financial guardrails within Snowflake to ensure cost efficiency, accountability, and operational discipline. This chapter is designed to help you take back control, intentionally, proactively, and with confidence. These strategies help organizations proactively manage cloud spending, optimize resource usage, and enforce financial policies.

During this phase, you explore how to establish financial guardrails that keep your Snowflake usage aligned with your budget and business goals. Think of it as setting up the rules of the road before you hit the gas.

As illustrated in Figure 4-1, the strategies covered in this chapter include:

- Monitoring usage and setting limits to prevent overspending
- Creating budgets that track spending across teams and projects
- Automating warehouse behavior with suspend/resume settings to eliminate idle costs
- Implementing chargeback and showback models to promote accountability
- Controlling access to ensure only the right people can perform cost-incurring actions

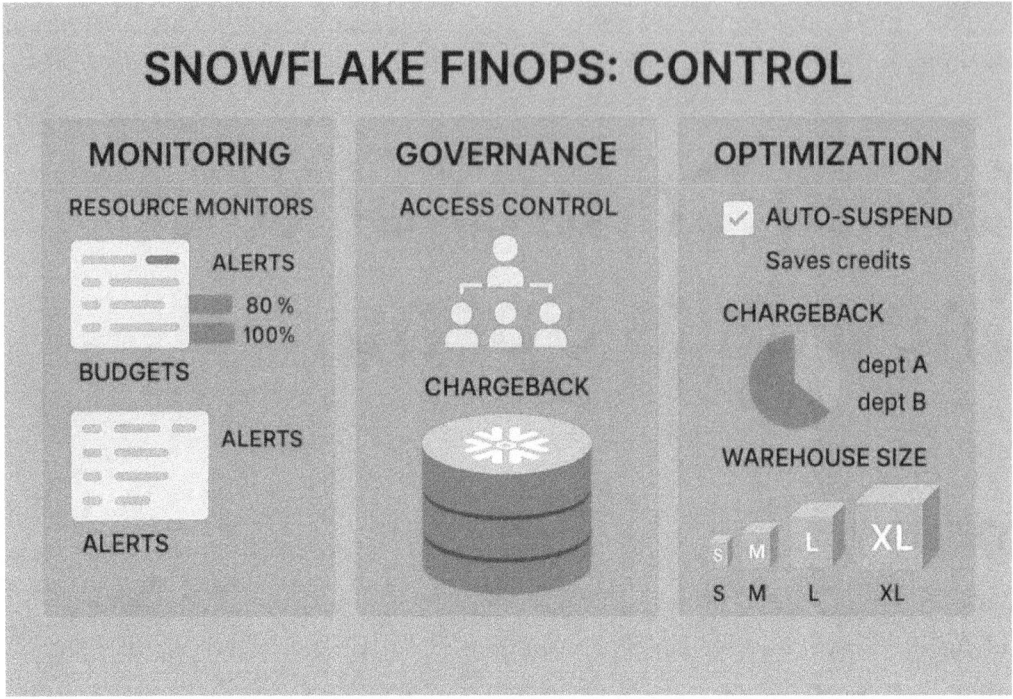

Figure 4-1. Snowflake FinOps Control phase

These are not just technical configurations. They are strategic tools that help you build a culture of financial discipline. By the end of this chapter, you will be equipped to manage your Snowflake environment with clarity and control. This chapter begins by exploring how to monitor usage and set effective limits to manage Snowflake costs proactively.

Monitoring Usage and Setting Limits: The Foundations of Cost Control

Snowflake offers unprecedented scalability and performance. However, this flexibility comes with a new kind of responsibility: managing consumption-based costs. Unlike traditional infrastructure, where resources are provisioned and paid for upfront, Snowflake operates on a pay-as-you-go model. This means that every query, every virtual warehouse, and every second of compute time has a direct financial implication.

To navigate this landscape effectively, organizations must adopt a proactive approach to monitoring and controlling usage. Snowflake provides native tools that make this possible, enabling platform admins to define usage boundaries, receive alerts, and take automated actions when thresholds are crossed. These tools are not just technical features; they are essential instruments of financial governance.

Resource Monitors and Credit Quotas

Resource monitors are configurable entities that track the consumption of Snowflake credits by virtual warehouses. Platform admins can define *credit quotas,* which are limits on how many credits can be used over a specified period, such as daily, monthly, or annually.

This mechanism serves multiple purposes. First and foremost, it acts as a safeguard against unexpected cost overruns. By setting clear boundaries, organizations can ensure that their usage aligns with budgetary expectations. Secondly, it encourages teams to be mindful of their resource consumption, fostering a culture of efficiency and accountability. Finally, it enables proactive intervention through alerts and automated actions, reducing the need for constant manual oversight.

For example, a marketing analytics team might be allocated a monthly budget of 5,000 credits. A resource monitor can be configured to send an alert when 90% of this quota is reached and to suspend the warehouse once 100% is consumed automatically. This setup ensures that the team remains within budget while still having the flexibility to operate independently.

Monitor Types and Granularity

Resource monitors can be applied at various levels, allowing platform admins to tailor their approach based on the needs of different teams or departments.

- **Account-level monitors** provide a holistic view of credit consumption across all virtual warehouses within a Snowflake account. This is ideal for centralized IT or finance teams that need to oversee total usage and enforce global policies.

- **Warehouse-level monitors**, on the other hand, offer more granular control. These monitors can be assigned to specific warehouses, making them suitable for departments or projects with distinct usage patterns and budgets.

As shown in Figure 4-2, this process demonstrates how to set up a resource monitor for the specified warehouse with a credit quota limit.

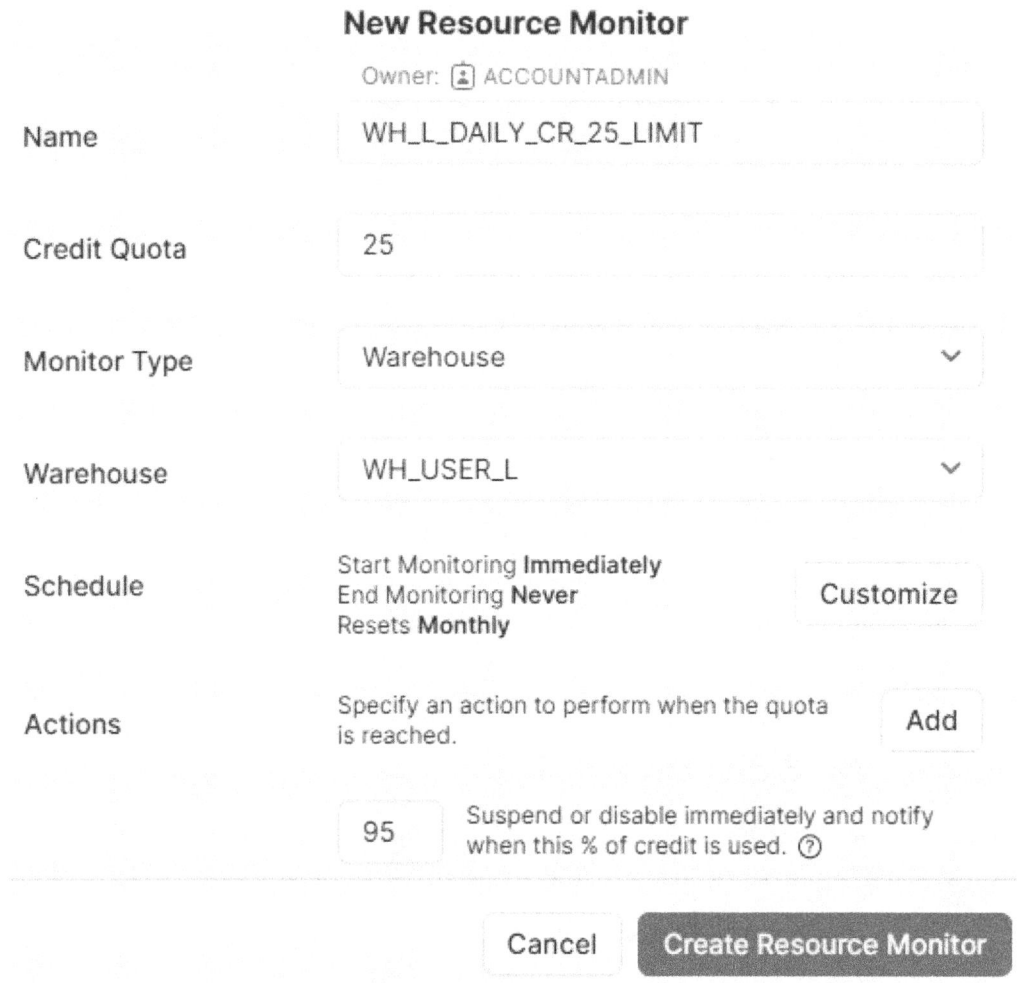

Figure 4-2. *Warehouse-level resource monitor setup*

This dual-layered approach allows for both broad oversight and targeted control, ensuring that no single team exceeds its allocated resources without visibility.

Automated Actions and Intelligent Alerts

Monitoring alone is not enough. What matters is how you respond when usage crosses a defined threshold. As illustrated in Figure 4-3, Snowflake allows you to configure specific actions that trigger automatically, ensuring timely and appropriate responses. These actions can be tailored to reflect the criticality of each environment and the organization tolerance for financial risk.

- **Notify:** Sends an alert to the designated user or stakeholders when a usage threshold is approached or exceeded.
- **Suspend:** Pauses the virtual warehouse after all currently running queries are completed.
- **Suspend immediately:** Terminates all running queries and halts the warehouse instantly.

This ensures that experimental or non-essential workloads do not consume more than their fair share of resources, while production environments remain unaffected.

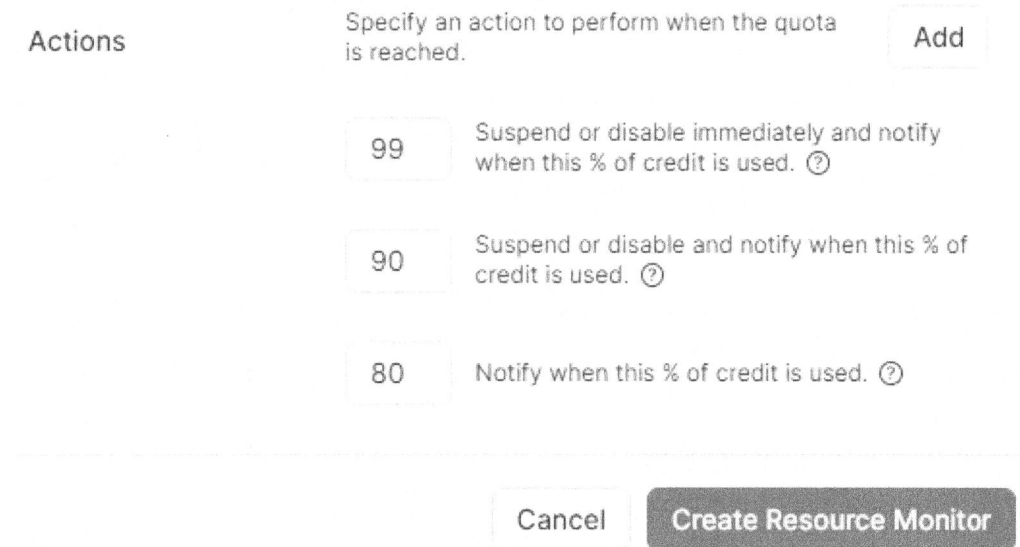

Figure 4-3. Resource monitor action setup

CHAPTER 4   TAKING CONTROL OF YOUR SNOWFLAKE SPEND

Best Practices for Sustainable Monitoring

Implementing resource monitors is not a one-time task; it is an ongoing process that requires regular review and adjustment, as shown in Figure 4-4. To ensure long-term effectiveness, organizations should adopt a set of best practices that align with their operational and financial goals. Consider these:

- **Review usage trends regularly:** Monitoring historical usage data helps identify patterns, forecast future needs, and adjust quotas accordingly. This prevents both underutilization and overconsumption.

- **Set realistic thresholds:** Quotas should be based on actual usage data rather than arbitrary limits. Overly restrictive quotas can lead to unnecessary disruptions, while overly generous ones may fail to enforce discipline.

- **Use auto-suspension strategically:** While auto-suspension is a powerful tool, it should be applied judiciously. It is best suited for non-critical environments where the risk of interruption is acceptable.

Cost Management

Organization Overview	Account Overview	Consumption	Anomalies PREVIEW	Budgets	Resource Monitors

3 Resource Monitors

NAME	QUOTA USED ↑	LEVEL	WAREHOUSES	FREQUENCY
WH_L_DAILY_CR_25_LIMIT	6.76%	Warehouse	WH_USER_L	Monthly
COMPUTE_WH_GT_10_CR	55.30%	Warehouse	COMPUTE_WH	Monthly
GEN2_DAILY_CR_10_LIMIT	92.80%	Warehouse	GEN2_WAREHOUSE +1	Monthly

Figure 4-4. Resource monitors quota usage

Common Pitfalls to Avoid

Before implementing a monitoring strategy, it is important to be aware of common pitfalls that can undermine its effectiveness.

- **Ignoring alerts:** Alerts are useful only if they prompt action. Failing to respond to notifications can result in budget overruns and operational surprises.

- **Setting quotas too low:** Unrealistically low limits can cause frequent interruptions, reducing productivity and frustrating users.

- **Neglecting to update monitors:** As usage patterns evolve, so too should the monitoring strategy. Static configurations can quickly become obsolete, leading to either lax control or excessive restrictions.

By following this structured approach, organizations can gain greater visibility into their Snowflake usage, enforce financial discipline, and create a more predictable and manageable cloud cost environment.

The next section explores the critical role that budgets play in effective cloud cost governance.

The Role of Budgets in Cloud Cost Governance

Budgets in Snowflake serve as predefined spending limits that can be applied across various scopes, entire accounts, specific departments, or individual projects. When paired with alerts, these budgets become powerful instruments for financial governance, enabling organizations to monitor usage, enforce accountability, and respond swiftly to potential overspending.

As illustrated in Figure 4-5, setting a monthly budget provides visibility into budget usage, remaining credits, and the number of days left in the current cycle.

CHAPTER 4 TAKING CONTROL OF YOUR SNOWFLAKE SPEND

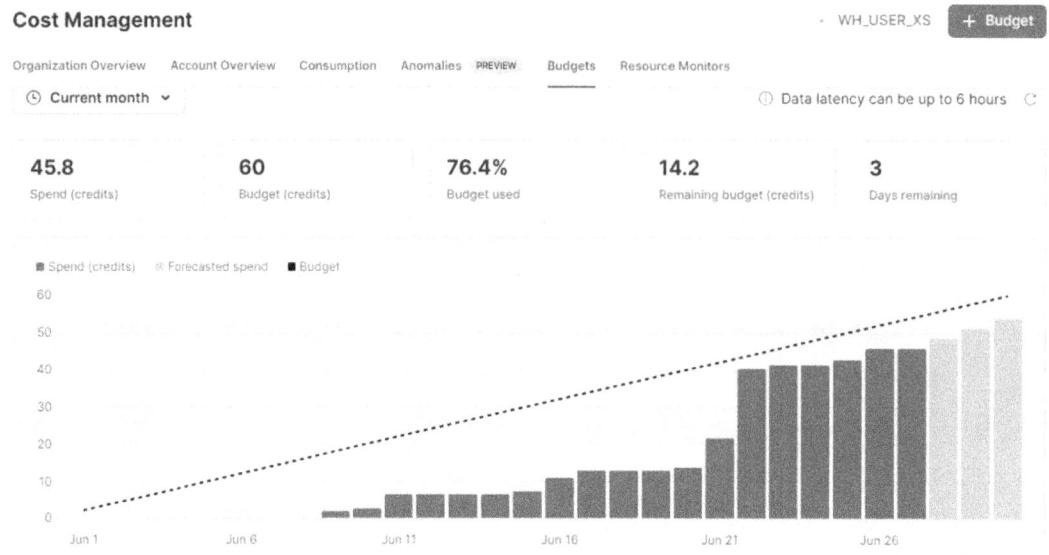

Figure 4-5. Setting up a budget on credit usage

Benefits of Budgeting and Alerts

Implementing budgets and alerts in Snowflake offers several strategic advantages:

- **Cost control:** Budgets help ensure that spending remains within acceptable limits, reducing the risk of unexpected charges.

- **Financial accountability:** By assigning budgets to specific teams or projects, organizations can track spending at a granular level and hold stakeholders accountable.

- **Proactive management:** Alerts provide early warnings when spending approaches or exceeds budget thresholds, allowing for timely interventions and adjustments.

These capabilities transform budgeting from a reactive process into a proactive discipline, aligning financial oversight with operational agility.

How Budgeting Works in Snowflake

The budgeting process in Snowflake involves three key components: defining budgets, configuring alerts, and monitoring usage.

- **Creating budgets:** Platform admins can define budgets based on organizational structure, such as by department, business unit, or project. Budgets can also be time-bound, aligning with fiscal periods or project timelines.

- **Setting alerts:** Alerts are configured to trigger when spending reaches specific thresholds, commonly at 80%, 90%, and 100% of the budget. These alerts can be sent to designated stakeholders, ensuring visibility and accountability.

- **Monitoring and reporting:** Snowflake provides built-in tools for tracking spending in real time and generating reports. These insights help organizations understand usage trends, evaluate budget performance, and make informed decisions.

Account Budgets: Organization-Wide Oversight

Account budgets provide a top-down view of spending across the entire Snowflake environment. These budgets encompass all supported objects and resources, offering a comprehensive perspective on total usage and cost.

- **Comprehensive spending limits:** Set a unified budget for the entire account, covering all compute and storage resources.

- **Holistic monitoring:** Track usage across all objects to gain a complete understanding of organizational spending.

- **Daily notifications:** Receive daily alerts if projected spending is expected to exceed the defined budget, enabling early intervention.

Custom Budgets: Granular Financial Management

While account budgets offer a broad view, custom budgets enable more targeted control. These budgets can be applied to specific resource groups (such as departments, projects, or business units), allowing for detailed oversight and tailored financial management.

- **Granular budgeting:** Define budgets for individual teams or initiatives, aligning financial limits with operational responsibilities.
- **Targeted monitoring:** Focus on specific Snowflake objects, including warehouses, databases, and user groups.
- **Custom alerts:** Configure alerts based on the unique needs of each budget, ensuring that relevant stakeholders are informed.

Spending Limits and Alerts: Enforcing Financial Discipline

In addition to budgets, Snowflake allows platform admins to define *spending limits*, which are hard caps on resource usage within a given time frame. These limits can be enforced through alerts and automated actions, providing a safety net against uncontrolled spending.

- **Defined spending limits:** Specify maximum allowable spending on Snowflake credits for a calendar month or custom period.
- **Proactive alerts:** Receive notifications when usage trends suggest that the limit may be exceeded.
- **Automatic suspension:** Optionally suspend resources when the limit is reached, preventing further charges.

Best Practices for Budget Management

To maximize the effectiveness of Snowflake budgeting and alerting tools, organizations should adopt the following best practices.

Regular monitoring:

- Review budgets frequently to ensure they reflect current usage patterns and business priorities.

- Adjust limits and alert thresholds as needed to accommodate growth or changes in workload.

Realistic budgeting:

- Base spending limits on historical usage data and forecasted needs.
- Avoid setting limits that are too restrictive or too lenient, as both can undermine the effectiveness of monitoring.

Strategic use of suspensions:

- Enable automatic suspension for non-critical environments to prevent unnecessary costs.
- Ensure that critical operations are protected by setting appropriate thresholds and exclusions.

Advanced Controls and Future Considerations

As organizations mature in their use of Snowflake, they may explore more advanced budgeting capabilities, such as these:

- **Custom alerts:** Developing tailored alerting mechanisms based on specific query patterns or usage metrics (e.g., query_hash, query_parameter_hash).
- **Anomaly detection:** Leveraging machine learning or third-party tools to detect unusual spending behavior. (See Chapter 13 for details.)

With budgets providing a strong foundation for financial oversight, the next step is to ensure accountability by attributing costs to the teams that generate them. The next section explores how chargeback and showback models support this goal.

Applying Chargeback and Showback Models

As organizations scale their use of cloud platforms like Snowflake, understanding and managing costs become increasingly complex. In a FinOps framework, visibility and accountability are essential. One of the most effective ways to achieve this is through the implementation of chargeback and showback models.

These models are designed to attribute cloud costs to the teams, departments, or business units that generate them. By doing so, they promote financial responsibility, encourage efficient resource usage, and align spending with organizational goals. Whether through direct billing or transparent reporting, chargeback and showback help transform cloud cost management from a centralized IT concern into a shared organizational responsibility.

Understanding Chargeback and Showback Models

At their core, chargeback and showback serve the same purpose: to make cloud spending visible and actionable. However, they differ in how they enforce accountability.

- **Chargeback** is a model in which individual teams or departments are billed directly for their usage of Snowflake resources. This creates a clear financial link between consumption and cost, encouraging teams to manage their usage efficiently.

- **Showback**, by contrast, provides detailed reports on usage and associated costs without actual billing. It raises awareness and fosters accountability, but without the immediate financial implications of a chargeback.

Both models are valuable tools in a FinOps strategy. Organizations often begin with showback to build awareness and gradually transition to chargeback as teams mature in their cost-management practices.

Implementing the Chargeback Model

The *chargeback model* is a direct and disciplined approach to cost attribution. It involves tracking resource usage at a granular level and billing each department or business unit accordingly. This model is particularly effective in organizations where financial accountability is a priority and where departments operate with independent budgets.

- **Direct billing:** Departments are charged based on their actual use of Snowflake resources.

- **Financial responsibility:** Teams are held accountable for their consumption, encouraging them to optimize usage.

- **Detailed reporting:** Usage data is collected and analyzed to support accurate billing and budget planning.

Implementing the Showback Model

The *showback model* offers a softer approach to cost attribution. Instead of billing departments directly, it provides them with detailed visibility into their usage and associated costs. This model is ideal for organizations that are new to FinOps or that want to build cost awareness without introducing financial pressure.

- **Usage visibility:** Departments receive reports detailing their resource consumption and related costs.

- **Awareness and accountability:** Teams gain insight into their usage patterns, encouraging more thoughtful resource management.

- **Non-intrusive oversight:** No direct billing is involved, making it easier to implement and manage.

Best Practices for Implementation

To ensure the success of the chargeback and showback models, organizations should follow this set of foundational best practices.

Ensure accurate tagging:

- Implement a consistent and comprehensive tagging strategy to attribute usage correctly.

- Regularly audit and update tags to reflect changes in organizational structure or resource allocation.

Communicate policies clearly:

- Clearly define and communicate the rules and expectations around chargeback and showback.

- Provide training and documentation to help teams understand how their usage is measured and reported.

Use showback as a transitional tool:

- Begin with showback to build awareness and prepare teams for financial accountability.
- Use showback data to identify trends, educate stakeholders, and lay the groundwork for a future chargeback model.

Leverage automation and reporting:

- Automate data collection and reporting to reduce administrative overhead.
- Use dashboards and scheduled reports to keep stakeholders informed and engaged.

Driving Financial Accountability Through Visibility

Chargeback and showback models are essential components of a mature FinOps strategy. By attributing costs to the teams that generate them, these models promote transparency, accountability, and efficiency. Whether implemented independently or as part of a phased approach, they help organizations align cloud spending with business priorities and foster a culture of cost-consciousness.

As Snowflake usage grows, so too does the importance of understanding who is using what, and at what cost. Chargeback and showback provide the tools to answer these questions and to manage cloud resources, not just as technical assets, but as financial ones.

After establishing financial accountability through chargeback and showback, the next step is to optimize operations. The next section looks at how automation can reduce idle costs and improve warehouse efficiency in Snowflake.

Automating Warehouse Efficiency

The Snowflake consumption-based pricing model rewards organizations that can align compute usage with actual demand. One of the most effective ways to achieve this alignment is through the use of the auto-suspend and auto-resume features for virtual warehouses.

These automation settings are designed to reduce idle time and eliminate unnecessary costs by pausing compute resources when they are not in use and restarting them only when needed. By leveraging these features, organizations can ensure that they are paying only for what they use, without sacrificing performance or availability.

Understanding Auto-Suspend and Auto-Resume

Snowflake warehouses are the engines that power query execution. When left running, they continue to consume credits, even if no queries are being processed. Auto-suspend and auto-resume provide a mechanism to manage this behavior intelligently.

- **Auto-suspend** automatically pauses a warehouse after a specified period of inactivity. This prevents the accumulation of charges for idle compute time.

- **Auto-resume** automatically restarts a suspended warehouse when a new query is issued, ensuring that users experience minimal disruption.

Together, these features form a powerful cost-control mechanism that operates silently in the background, optimizing resource usage without requiring manual oversight.

Key Benefits of Automation

Implementing auto-suspend and auto-resume offers several operational and financial advantages:

- **Cost efficiency:** Idle warehouses are suspended automatically, ensuring that you are not billed for unused compute time.

- **Resource optimization:** Compute resources are made available precisely when needed and paused when not, maximizing utilization.

- **Operational simplicity:** Automation reduces the need for manual monitoring and intervention, freeing up administrative resources.

These benefits make auto-suspend and auto-resume essential tools for any organization seeking to implement a disciplined and efficient FinOps strategy.

> **Note** For a deeper exploration of warehouse scaling policies and metrics to monitor, refer to Chapter 5.

Combine with Budgeting and Alerts

To get the most out of Snowflake's cost-saving features, it is helpful to pair auto-suspend and auto-resume with your budgeting and alerting tools. This combination gives you better control over both usage and spending.

For instance, if a warehouse is resuming frequently, it could mean the inactivity timeout is too short. Adjusting it can reduce unnecessary restarts and save credits. By monitoring these patterns and setting up alerts, you can fine-tune warehouse behavior to better align with your budget goals, all while maintaining performance.

While automation helps eliminate waste, optimizing performance requires a more strategic approach. The next section explores how to balance compute power and cost through effective warehouse sizing.

Balancing Performance and Cost

One of the most impactful decisions is selecting the right size for your compute resources. In Snowflake, this means choosing the appropriate virtual warehouse size to match your workload requirements. The goal is to strike a balance: provision enough compute power to ensure performance, but not so much that you incur unnecessary costs.

Snowflake offers a range of warehouse sizes, each with distinct compute capabilities and cost implications. Understanding how to select, scale, and manage these sizes is essential for optimizing both performance and expenditure.

What Are Warehouse Sizes and Why Do They Matter?

A *Snowflake warehouse* is a compute cluster used to execute queries, load data, and perform transformations. These warehouses come in predefined sizes, ranging from X-Small to 6X-Large, with each size doubling in compute capacity and cost as you move up the scale.

The choice of an appropriate warehouse size impacts not only performance but also cost. Over-provisioning leads to wasted spend, while under-provisioning can degrade performance and delay results. Snowflake's flexible architecture allows you to resize warehouses as needed, making it possible to adapt to changing workloads without long-term commitments.

Key Benefits of Right-Sizing Warehouses

Selecting the appropriate warehouse size offers several strategic advantages:

- **Optimized performance:** Ensures that queries and operations are executed efficiently, reducing wait times and improving user experience.

- **Cost management:** Aligns compute capacity with actual workload needs, avoiding the financial waste of over-provisioning.

- **Scalability:** Enables dynamic scaling to accommodate workload spikes or seasonal demand, without overhauling infrastructure.

These benefits make warehouse sizing a foundational element of cost-conscious Snowflake operations.

How Warehouse Sizing Works

The Snowflake warehouse sizing model is designed for simplicity and flexibility:

- **Warehouse sizes:** Available sizes include X-Small, Small, Medium, Large, X-Large, 2X-Large, 3X-Large, up to 6X-Large. Each step up doubles the compute resources and associated cost.

- **Auto-scaling:** Warehouses can be configured to automatically add clusters during periods of high concurrency, ensuring consistent performance.

- **Concurrency and limits:** Larger warehouses support more concurrent queries and faster processing, but also consume more credits per second.

This model allows organizations to tailor compute resources to the specific needs of each workload, adjusting as those needs evolve.

Warehouse Sizing: Making the Right Choice

Choosing the right warehouse size requires a thoughtful assessment of workload characteristics, including data volume, query complexity, and concurrency requirements.

- **Range of sizes:** Sizes range from lightweight X-Small warehouses for development and testing to powerful 6X-Large clusters for enterprise-scale analytics.
- **Compute resources:** Each size includes a defined amount of CPU and memory, directly impacting processing speed and throughput.
- **Scalability:** Warehouses can be resized manually or automatically, allowing for responsive adjustments to workload demands.

Best Practices for Warehouse Sizing

To make the most of the Snowflake flexible sizing model, consider the following best practices.

Match size to workload:

- Use smaller warehouses for lightweight, low-concurrency tasks.
- Reserve larger sizes for high-volume, high-concurrency operations or time-sensitive processing.

Monitor and adjust regularly:

- Analyze query performance and warehouse utilization metrics to identify inefficiencies.
- Resize warehouses based on observed patterns rather than assumptions.

Leverage auto-scaling:

- Enable multi-cluster auto-scaling for workloads with unpredictable concurrency.
- Set appropriate limits to avoid runaway costs while maintaining performance.

Avoid static sizing:

- Don't treat warehouse size as a one-time decision. Reevaluate regularly as workloads evolve.

Reflection questions:

- How can you determine the appropriate warehouse size for different types of workloads in your Snowflake environment?
- What metrics or usage patterns would you monitor to decide when to resize a warehouse?
- How might auto-scaling complement your warehouse sizing strategy to balance cost and performance?

Beyond managing compute and cost, true financial governance also requires controlling who can take cost-incurring actions. The next section examines how access control plays a vital role in FinOps.

The Role of Access Control in FinOps

In the context of FinOps, cost control is not limited to compute usage and storage optimization, it also extends to those who have access to what. Every action in Snowflake, from running a query to creating a warehouse, has a cost implication. Therefore, managing access is not only a matter of security and compliance but also a critical component of financial governance.

Snowflake provides a robust and flexible access control framework that allows organizations to define, enforce, and audit permissions across users and roles. By implementing structured access control policies, organizations can prevent unauthorized usage, reduce the risk of cost overruns, and ensure that only the right people have the ability to perform cost-incurring operations.

Why Access Control Matters

Effective access control is foundational to both data security and operational efficiency. It ensures that sensitive data is protected, regulatory requirements are met, and users can perform their duties without unnecessary privileges.

- **Data security:** Prevents unauthorized access to sensitive or regulated data.
- **Compliance:** Supports adherence to industry standards and internal governance policies.
- **Operational efficiency:** Simplifies the management of user permissions, reducing administrative overhead and risk.

How Access Control Works in Snowflake

Snowflake access control model is built around *role-based access control* (RBAC), with support for granular permissions and auditing capabilities. These components work together to provide a secure and manageable framework for controlling access.

- **RBAC:** Permissions are assigned to roles, and users are assigned to those roles. This abstraction simplifies permission management and enhances scalability.
- **Granular permissions:** Access can be controlled at the level of specific actions (e.g., SELECT, INSERT) and resources (e.g., databases, tables).
- **Audit and monitoring:** Snowflake logs user activity and access events, enabling organizations to monitor usage and detect anomalies.

Role-Based Access Control (RBAC): A Scalable Model

RBAC is the cornerstone of Snowflake's access control system. It allows platform admins to define roles based on job functions and assign permissions to those roles. Users inherit permissions through their assigned roles, making it easier to manage access across large and dynamic teams.

- **Role assignment:** Permissions are granted to roles, not directly to users.

- **User mapping:** Users are assigned to one or more roles, inheriting the associated permissions.

- **Role hierarchies:** Roles can be nested, allowing for inheritance and delegation of permissions.

A financial institution defines roles such as data analyst, data engineer, and admin. Each role is granted specific permissions aligned with its responsibilities. By assigning users to these roles, the organization ensures that only authorized personnel can access or modify sensitive financial data.

Granular Permissions: Precision in Access Control

Beyond role assignment, Snowflake allows for fine-grained control over which actions users can perform and on which resources. This level of precision is essential for enforcing the *principle of least privilege,* which is defined as granting users only the access they need to perform their tasks.

- **Action-specific permissions:** Control over operations such as SELECT, INSERT, UPDATE, DELETE, and EXECUTE.

- **Resource-specific permissions:** Permissions can be scoped to specific databases, schemas, tables, views, and more.

- **Custom role design:** Roles can be tailored to match the exact needs of different teams or projects.

For example, in a financial institution, investment analysts are granted read-only access to client portfolios and market data, while portfolio managers have write access to adjust asset allocations. This ensures that sensitive financial data is protected while still enabling authorized personnel to perform critical investment operations.

The Principle of Least Privilege and Regular Permission Reviews

As mentioned, adhering to the principle of least privilege means granting users the minimum level of access required to perform their duties. This principle is a cornerstone of both security and cost control.

- **Regular permission reviews:** Periodically audit user roles and permissions to ensure they remain appropriate.
- **Access revocation:** Remove unnecessary or outdated permissions to reduce risk.
- **Change management:** Update roles and permissions as users change roles or projects.

For example, a retail company uses Snowflake access logs and audit reports to monitor who accesses customer data. Alerts are configured to flag unusual access patterns, and permissions are reviewed quarterly to ensure alignment with current job roles.

Monitoring and Auditing: Visibility into Access

Snowflake provides built-in tools for activity logging, access auditing, and alerting, which are essential for maintaining oversight and responding to potential security incidents.

- **Activity logging:** Track user actions, including logins, queries, and data modifications.
- **Audit trails:** Maintain detailed records of access to sensitive data and administrative operations.
- **Alerts and reports:** Enable real-time alerts and scheduled reports for compliance and security monitoring.

Best Practices for Access Control in Snowflake

To implement effective access control, organizations should adopt the following best practices.

Establish a clear access control policy:

- Define roles and responsibilities clearly.
- Document permission requirements for each role.

Implement RBAC thoughtfully:

- Create roles based on job functions, not on individuals.
- Use role hierarchies to simplify management.

Conduct regular access reviews:

- Schedule periodic audits of user roles and permissions.
- Adjust access as team structures and responsibilities evolve.

Monitor and audit continuously:

- Enable logging and alerting for sensitive operations.
- Review audit logs regularly to detect anomalies.

Reflection questions:

- How could implementing RBAC in your Snowflake environment improve security and simplify permission management?
- What roles would you define, and what permissions would they include?
- How can regular permission reviews and audit logs help you enforce the principle of least privilege?

Summary

The Control phase in FinOps empowers organizations to manage Snowflake usage with discipline and clarity. By leveraging monitors, budgets, alerts, and automation, teams gain visibility, prevent overspending, and reduce idle compute time. Smart warehouse sizing and role-based access controls further enhance efficiency and security. Together, these practices lay a strong foundation for financial accountability and prepare teams to move into the Optimization phase, where the focus shifts to continuous improvement and long-term value.

In the upcoming chapter, we shift the focus to compute optimization, where the goal goes beyond managing resources to continuously improving how they are allocated and consumed. We explore strategies for tuning warehouse performance, eliminating inefficiencies, and maximizing the value of every Snowflake credit.

CHAPTER 5

Compute Optimization Strategies

Overview

Optimizing compute resources in Snowflake not just about allocating compute (scale up and scale out) when needed. It is about adopting a proactive and continuous approach of resource management. The key is balancing performance with cost by fine-tuning how compute is allocated as your workloads evolve. Instead of occasional warehouse tweaks, the compute optimization process of Snowflake requires ongoing adjustments to workload architecture to ensure efficiency.

This chapter covers best practices for configuring virtual warehouses (VWs), eliminating resource waste, right-sizing compute capacity, and measuring utilization. But you dive into the practical strategies that can lead to significant cost savings, it is worth orienting yourself to Snowflake's internal architecture, particularly its warehouse cluster-level MPP (massively parallel processing) design. This architecture is the foundation of how compute resources are consumed, so grasping it fully will set the stage for applying the strategies that follow. We break it down and show you how it ties into real-world scenarios, complete with practical examples and scripts to help you optimize your Snowflake environment efficiently.

The essential compute optimization strategies covered in this chapter include:

- Adjusting warehouse size based on utilization trends to balance cost and performance.

- Optimizing idle time settings to save money while maintaining query performance.

- Dynamically scaling clusters using Standard, Economy, or Maximized mode based on workload demand.

- Boosting compute power for resource-intensive queries on-demand without resizing warehouses.

- Combining similar workloads into fewer warehouses to reduce resource waste and improve efficiency.

- Using Snowflake's utilization views to analyze warehouse efficiency and make informed adjustments.

These strategies help optimize compute resources, reduce costs, and ensure high performance in Snowflake environments.

MPP Compute: The Role of Parallelism

When we provision a virtual warehouse in Snowflake, it seems like a single physical compute system with a preferred size (e.g., Medium) is provisioned from a high-level perspective. However, beneath the surface, there is actually a cluster of physical compute layers working together as a single cluster, which is presented as one virtual warehouse (VW). The warehouse provides the required resources, such as CPU, memory, and temporary storage, to perform SELECT, UPDATE, DELETE, and INSERT commands. When a user submits a request to read or write a table, Snowflake breaks that request into parallel threads that are executed simultaneously across the nodes.

Unlike other MPP database systems, Snowflake achieves its best performance when parallelism is fully utilized. To optimize this, you need to start by carefully sizing the files used for data loading, selecting the right cluster key for large tables, and avoiding common pitfalls like join explosion. Join explosion can lead to parallelism inefficiencies, causing the requests to take longer than expected.

As depicted in Figure 5-1, in an MPP system, the execution of each query is distributed across all available processors. This architecture leverages the collective computational power to efficiently handle substantial volumes of data. More processors lead to better performance and higher throughput.

Additionally, reducing unnecessary network data movement and ensuring that all nodes in the cluster are used efficiently are critical for achieving optimal performance. By focusing on these factors, you can ensure that Snowflake's parallel processing capabilities are fully leveraged, resulting in faster and more efficient queries.

CHAPTER 5 COMPUTE OPTIMIZATION STRATEGIES

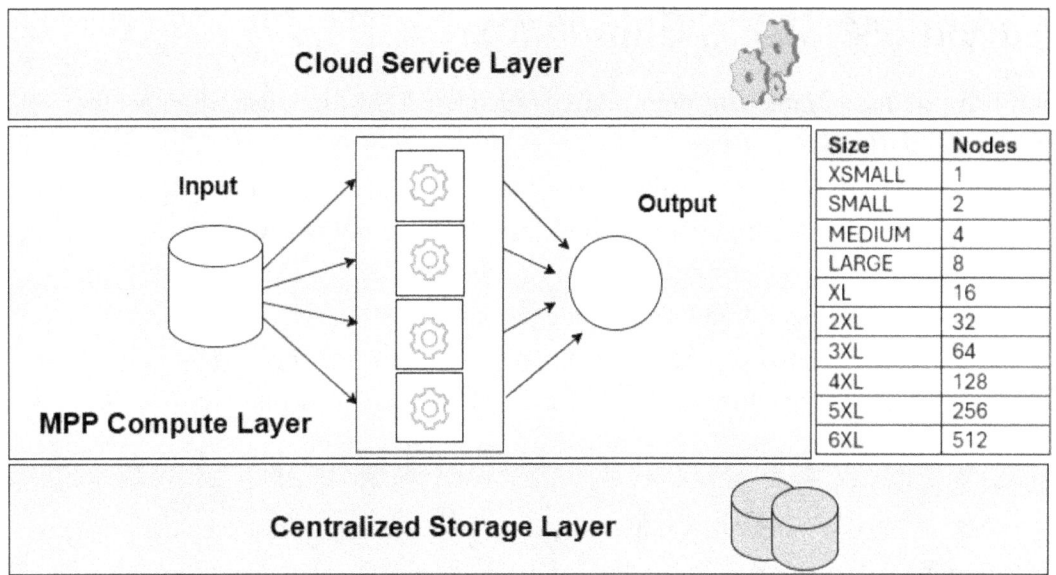

Figure 5-1. *Snowflake MPP compute layer*

On the other hand, in consumption-based resource provisioning, finding the right balance is the key to efficiency. Over-provisioning leads to wasted resources and unnecessary costs, while under-provisioning can result in poor performance that negatively impacts your business. The trick is to find that sweet spot, the median point for optimal resource handling.

Database systems, typically built on a capital expenditure (CapEx) model, have traditionally focused on throughput and performance. To handle peak loads and ensure flexibility, they often rely on over-provisioning resources, which can result in inefficiencies and wasted capacity. While this approach may work in certain situations, it's not the most cost-effective solution. Snowflake offers a different model to meet capacity demand only when required, charging only for the time you use the resources. With virtually unlimited compute and storage resources, Snowflake shifts the focus from just managing performance to effectively managing costs.

While you get the required compute resources at a click of a button, it's important to size the virtual warehouse to meet the demand, while over-provisioning ends up in waste.

CHAPTER 5   COMPUTE OPTIMIZATION STRATEGIES

Warehouse Sizing Guidelines

Why use and pay for large compute when a much smaller warehouse will meet the desired service levels?

When utilization is consistently low over several days (for example, < 10), and there is no spillage to remote storage for the queries executing in those warehouses, consider downsizing the warehouse or reducing the number of clusters. This might reduce warehouse cost while minimally impacting performance.

When there is spillage to remote or local storage, consider upsizing the warehouse to improve performance. In some cases, this might be more cost effective (if spillage is avoided in an upsized warehouse, the performance gains might more than offset the cost of the larger warehouse).

When sizing the warehouse, make sure to choose the appropriate scaling policy to save money and resources.

Adjust Auto-suspend Settings

When the execution time of the initial query after a warehouse resumes is not critical, the auto-suspend property can be set to a low value (for example, 60 seconds) to optimize costs. Building the warehouse cache after resuming the warehouse leads to delay for initial queries when you set a low time period for auto-suspend.

On the other hand, when the query performance is important and the warehouse has frequent queries (for example, queries for a user-facing, latency-sensitive application), then the auto-suspend property can be set to a few minutes (for example, 2-5 minutes) to avoid frequent warehouse suspensions and the loss of cache.

Lowering the auto-suspend limits can result in significant credit savings, as Snowflake charges for every second of usage. The auto-suspend feature applies to the last active cluster; if multiple clusters are running, the additional clusters are shut down according to the scaling policy.

Pros of lowering auto-suspend:

- Can help reduce the credits used by the warehouse, since the warehouse suspends sooner when it's idle.

Cons of lowering auto-suspend:

- The warehouse cache is lost when a warehouse suspends, so suspending it too quickly may hurt ad hoc query performance.

- Suspending a warehouse too quickly on a busy system may cause it to start and stop more frequently. Every time the warehouse starts, there is one-minute charge, so lowering the threshold in this scenario can actually increase credit use.

- When a warehouse shuts down, the first new query using a warehouse may experience a delay while the warehouse starts up.

An important consideration when enabling the scaling policy is to understand Snowflake's billing system for virtual warehouses. When your application executes short queries that complete in a few seconds, provisioning a new cluster will incur a minimum billing period of 60 seconds (one minute). Selecting the Economy or Standard scaling policy can help mitigate these costs.

Adjust Multi-cluster Warehouse Settings

Virtual warehouses (VWs) with multiple clusters are referred to as multi-cluster VWs. As illustrated in Figure 5-2, each cluster consists of one or more servers, which are essentially virtual machines equipped with multiple vCPUs and local storage to facilitate processing and caching. This setup is comparable to an EC2 instance in Amazon Web Services. Snowflake leverages the offerings provided by cloud vendors to ensure consistent performance.

CHAPTER 5   COMPUTE OPTIMIZATION STRATEGIES

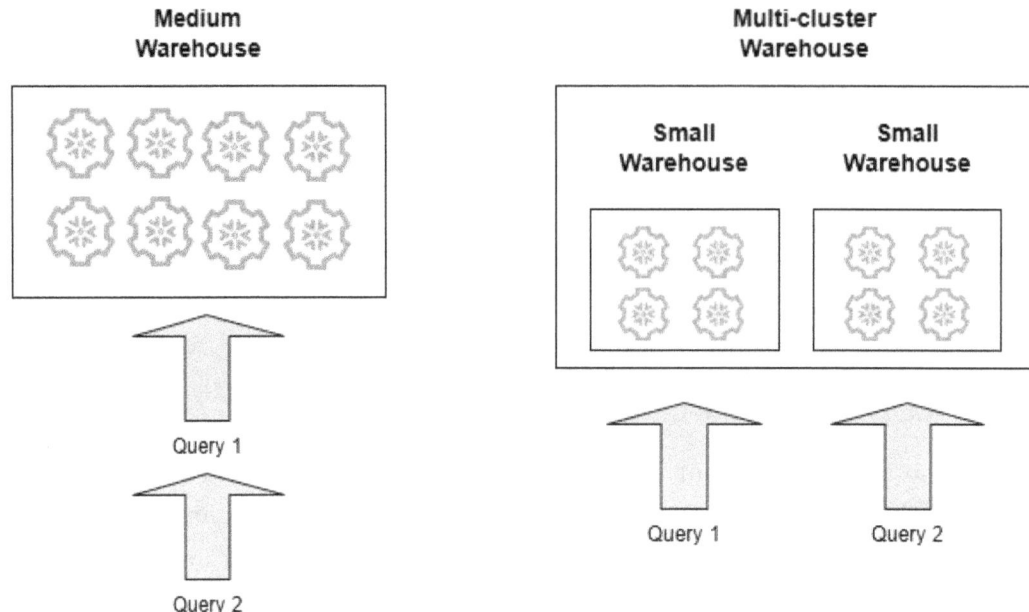

Figure 5-2. Snowflake multi-cluster warehouse

Auto-scaling dynamically adjusts the number of clusters based on the concurrent query load. When demand increases, Snowflake provisions additional clusters to manage the workload. Conversely, when demand decreases, clusters are suspended. It is important to note the 60-second minimum billing period: when a multi-cluster warehouse resumes, a minimum of 60 seconds of billing is incurred for each active cluster.

Snowflake offers two modes for auto-scaling—Economy and Standard—along with a third mode for maximum capacity when the warehouse is resumed. Selecting the appropriate policy allows you to meet your application's demand and response timeline effectively.

Standard Policy

The Standard policy for auto-scaling in Snowflake ensures efficient handling of query loads by dynamically adjusting the number of clusters based on demand. This approach aims to balance performance and cost-effectiveness.

> **Scale up:** The first additional cluster starts immediately when either a query is queued or the system detects that there's one more query than the currently-running clusters can execute.

Scale down: This happens after two to three consecutive successful checks (performed at one-minute intervals) to determine whether the load on the least-loaded cluster could be redistributed to the other clusters without spinning up the cluster again.

In this setup, the queries start and finish within the same cluster.

Pros:

- Since the system adds clusters as needed, there is reduced likelihood of queries being queued, minimizing user wait times for query initiation.

Cons:

- This approach can be slightly more expensive, as additional clusters may run for extended periods.

Economy Policy

The Economy policy for auto-scaling in Snowflake is designed to optimize cost savings by scaling clusters more conservatively based on query load.

Scale up: The system initiates an additional cluster when it estimates that the query load will keep the cluster busy for at least six minutes. This estimation includes both queued and recently completed queries.

Scale down: After five to six consecutive successful checks (performed at one-minute intervals) to determine whether the load on the least-loaded cluster can be redistributed to the other clusters without necessitating the reactivation of the cluster, the least-loaded cluster is scaled down and its workload, if any, is redistributed.

Pros:

- This option can be more cost-effective, typically saving around 5-10%, as it does not scale up as quickly as the Standard policy.

Cons:

- Users may experience longer wait times for queries due to more frequent queuing.

- Scaling down takes longer, which can reduce some of the cost savings.

- If there are insufficient queries to trigger the provisioning of a new cluster, smaller queries in the queue may experience significant delays before execution.

Maximized Mode

In Maximized mode, the number of clusters specified will be consistently running at all times. This mode is particularly advantageous when full control over compute resources is required, such as when managing a large number of concurrent users or handling queries with relatively stable volumes. It is ideal for scenarios where consistent performance and predictable resource allocation are essential.

To determine the appropriate scaling policy for multi-cluster warehouses, you can utilize the utilization metrics from the WAREHOUSE_CLUSTER_UTILIZATION or WAREHOUSE_UTILIZATION view:

- If the median utilization is low, the workloads are scheduled, and the queuing is acceptable, changing the scaling policy to economy might significantly improve utilization.

- If utilization is high (for example, > 70%) and the scaling policy is set to Standard, then Snowflake recommends leaving it set to Standard. Changing the scaling policy to Economy might not significantly improve utilization and could result in increased queuing.

While the scaling policy controls the duration before provisioning a new node to the cluster, it is equally important to monitor the number of sessions running on the cluster. When executing memory-intensive queries, reducing the number of concurrent sessions can help balance resource allocation and ensure optimal performance.

MAX_CONCURRENCY_LEVEL

In addition to scaling policies, another critical parameter to consider is MAX_CONCURRENCY_LEVEL. This parameter does not limit the number of statements that can be executed concurrently by a warehouse cluster. Instead, it serves as an upper boundary to prevent over-allocation of resources. By default, the concurrency level is set to 8.

This parameter essentially sets a cap on the number of full-slot queries that can run in parallel across all available vCPUs within a single cluster. However, it is not solely about the number of queries; memory usage is also considered when determining if a query can be added to the cluster. Each query receives a memory estimate, and the scheduler monitors the total memory being utilized by all active queries in the cluster. If adding a new query would exceed the defined memory limit, that query will not be dispatched, even if the number of full-slot queries is still below the MAX_CONCURRENCY_LEVEL. Instead, the query will remain in the queue.

Before adjusting this parameter, it is crucial to monitor factors such as virtual warehouse disk spill and any instances of running out of memory, as outlined in Table 5-1.

Table 5-1. *Pros and Cons of Adjusting Concurrency Levels*

Pros	Cons
Increasing concurrency levels: If the warehouse handles many small queries, raising the concurrency level lets more queries run at the same time on a cluster.	Increasing concurrency levels: Setting the concurrency level too high can cause queries to run out of memory (OOM), which is explained later in this chapter.
Decreasing concurrency levels: If the warehouse handles large, complex queries, lowering the concurrency level can help these queries run faster and prevent out-of-memory (OOM) issues.	Decreasing concurrency levels: Setting the concurrency level too low can cause new clusters to be created unnecessarily, even when the current cluster still has available bandwidth.

Based on the pros and cons outlined here, consider these specific recommendations:

- **Warehouse running primarily simple queries:** This can be increased to 12, 16, 20, 24. Always increase incrementally, watching for OOM errors.

- **Warehouse running large, complex queries:** Lowering this to 6 or 4 could improve the performance of those queries.

Next, queries that run for extended periods without any checks and controls can lead to excessive resource use. You can manage runaway sessions using account parameters, as outlined next.

STATEMENT_TIMEOUT_IN_SECONDS

The STATEMENT_TIMEOUT_IN_SECONDS parameter helps to control how long queries are allowed to run before they automatically get a timeout error. You can set it at the session, warehouse, or account level, depending on where you need control over query execution times.

This parameter does not just track the execution time of a query in the warehouse; it includes the total time the query takes, such as queue time, locked time, execution time, compilation time, and more. Essentially, it is the full time from when the query starts to when it finishes.

This timeout also applies to queries that do not require a warehouse to execute, such as CLONE, CREATE, or similar statements. Therefore, it provides a comprehensive way to manage query session times.

Recommendation:

1. Set the timeout between 30 to 60 minutes for ad hoc user sessions.

2. This can help prevent runaway queries, ensuring they do not run longer than necessary.

Lastly, the next section explores how the Query Acceleration Service (QAS) can provide additional compute resources to meet dynamic, ad hoc workload without scaling up the warehouse, which would leave compute resources unused when demand is low.

Query Acceleration Service (QAS) for Added Compute

The QAS allows Snowflake to automatically boost the compute power for a query on-demand, without needing any manual intervention. This feature is designed to be serverless, meaning it can handle the scaling on its own.

This service is especially useful when you have dynamic, ad hoc queries that might take longer to process, depending on the size of your data warehouse. In other words, as illustrated in Figure 5-3, if most of your queries (about 80%) run smoothly on the current warehouse size, the QAS comes into play for those 20% of queries that could benefit from a little extra compute power. Instead of resizing the entire warehouse, the service provides that additional capacity only when needed, saving time and resources.

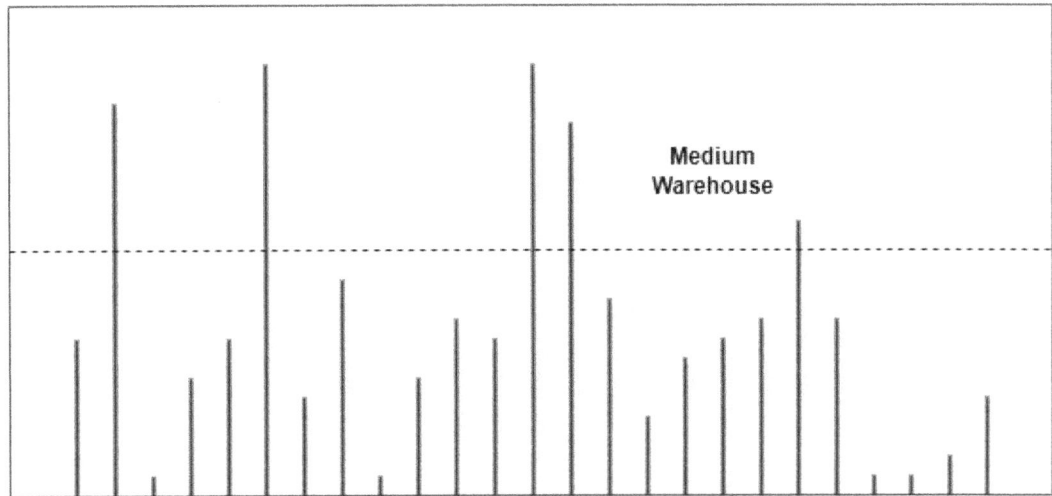

Figure 5-3. Query workload by hour

The QAS can be turned on at the warehouse level, and it works by automatically offloading certain parts of a query—usually long-running tasks like table scans and filters—to additional serverless compute resources. This not only enhances performance but also ensures cost efficiency, as you only pay for the additional compute when necessary. When you enable this service, set a scale factor, which lets you control how much extra compute power can be added to the base size of the warehouse.

Like any powerful tool, the QAS can lead to unexpectedly high costs if not used carefully. It's not recommended to enable it on every virtual warehouse.

For instance, if you run a query on a QAS-enabled Large warehouse and accidentally include a Cartesian join without any checks or controls in place, you might end up with much higher costs than you'd expect for a Large warehouse. Using QAS on a warehouse allows the warehouse to increase its compute by up to eight times its size when necessary.

CHAPTER 5   COMPUTE OPTIMIZATION STRATEGIES

To avoid runaway queries, you can set an automatic timeout for your user session or warehouse by configuring the STATEMENT_TIMEOUT_IN_SECONDS setting. Just be cautious; use Snowflake Resource Monitor to keep things in check.

Find Appropriate Queries

It is good to determine if the warehouse you are thinking of adding QAS to will benefit from it or not. WHAT | IF functionality indicates eligible queries.

Use the following SQL to identify queries that are eligible for the QAS.

```
SELECT
  QUERY_ID,
  QUERY_TEXT,
  WAREHOUSE_NAME,
  ELIGIBLE_QUERY_ACCELERATION_TIME,
  UPPER_LIMIT_SCALE_FACTOR
FROM SNOWFLAKE.ACCOUNT_USAGE.QUERY_ACCELERATION_ELIGIBLE
WHERE start_time > DATEADD('day', -7, CURRENT_TIMESTAMP())
ORDER BY ELIGIBLE_QUERY_ACCELERATION_TIME DESC;
```

To determine if the specific query will benefit from the QAS, run this query:

```
SELECT PARSE_JSON(SYSTEM$ESTIMATE_QUERY_ACCELERATION('8cd54bf0-1651-5b1c-ac9c-6a9582ebd20f'));
{
  "estimatedQueryTimes": {
    "1": 172,
    "10": 116,
    "2": 152,
    "4": 134,
    "8": 120
  },
  "originalQueryTime": 300.271,
  "queryUUID": "8cd53bf1-1661-5b1d-ak9c-6a9382ebd20f",
  "status": "eligible",
  "upperLimitScaleFactor": 10
}
```

In this example query, the original query time was 300.271 minutes and is significantly reduced by utilizing QAS.

If it is not eligible, the function will return ineligible status:

```
{
  "estimatedQueryTimes": {},
  "originalQueryTime": 31.291,
  "queryUUID": "c8d53bf1-1661-5b1d-ak9c-6a2382ebd20k",
  "status": "ineligible",
  "upperLimitScaleFactor": 0
}
```

Enable Acceleration Service to a Warehouse

Enabling QAS on a warehouse requires setting a simple parameter. You do this either during warehouse creation or by altering an existing warehouse:

```
CREATE WAREHOUSE WH_USER_S WITH
  ENABLE_QUERY_ACCELERATION = TRUE;
```

If you set the scale factor to 2, QAS can only lease four times as much compute as the original warehouse size.

```
ALTER WAREHOUSE WH_USER_S SET
  ENABLE_QUERY_ACCELERATION = TRUE
  QUERY_ACCELERATION_MAX_SCALE_FACTOR = 2;
```

Use QUERY_HISTORY to view QAS results of all queries at once.

```
SELECT query_id,
    query_text,
    warehouse_name,
    start_time,
    end_time,
    query_acceleration_bytes_scanned,
    query_acceleration_partitions_scanned,
    query_acceleration_upper_limit_scale_factor
FROM SNOWFLAKE.ACCOUNT_USAGE.QUERY_HISTORY;
```

To see how QAS affects your costs, check out the QUERY_ACCELERATION_HISTORY view:

```
SELECT warehouse_name,
    SUM(credits_used) AS total_credits_used
FROM SNOWFLAKE.ACCOUNT_USAGE.QUERY_ACCELERATION_HISTORY
WHERE start_time >= DATE_TRUNC(month, CURRENT_DATE)
GROUP BY warehouse_name
ORDER BY total_credits_used DESC;
```

While we follow warehouse sizing guidelines to meet workload demand, gaining in-depth insights into resource usage versus waste helps us take corrective action to maximize the investment in compute resources.

Use Data Scans for Sizing

There is another useful metric used for warehouse sizing: the number of micro-partitions a query scans. Based on that, you can select the appropriate warehouse size for the query. *This approach was shared by Scott Redding, a solutions architect at Snowflake.*

The core idea here is that as you increase the warehouse size, the number of available threads doubles. Each of these threads processes a single micro-partition at a time, and you want to make sure each thread has enough work to do throughout the query.

To understand this better, aim for about 250 micro-partitions per thread. For example, if a query needs to scan 2000 micro-partitions, running it on an X-Small warehouse will allocate 250 micro-partitions to each of its threads, which is optimal. On the other hand, running the same query on a 3-XL warehouse, which has 512 threads, would only give each thread four micro-partitions to process. This would likely leave many threads idle, which is not ideal for efficient execution.

In addition to the number of micro-partitions referenced in Figure 5-4, factors such as query complexity, join size, and the volume of sorted data also significantly influence the required processing power.

# of MPs / VWH	XS (8)	S(16)	M (32)	L(64)	XL (128)	2XL (256)	3XL (512)	4XL (1024)	5XL (2048)	6XL (4096)
2,000	250	125	63	31	16	8	4	2	1	0
4,000	500	250	125	63	31	16	8	4	2	1
8,000	1000	500	250	125	63	31	16	8	4	2
16,000	2000	1000	500	250	125	63	31	16	8	4
32,000	4000	2000	1000	500	250	125	63	31	16	8
64,000	8000	4000	2000	1000	500	250	125	63	31	16
128,000	16000	8000	4000	2000	1000	500	250	125	63	31
256,000	32000	16000	8000	4000	2000	1000	500	250	125	63
512,000	64000	32000	16000	8000	4000	2000	1000	500	250	125
1,024,000	128000	64000	32000	16000	8000	4000	2000	1000	500	250

Figure 5-4. Warehouse sizing based on partitions scanned (image courtesy Snowflake Inc.)

Metrics to Monitor

As shown in Table 5-2, this section reviews the metrics and guidelines you can use to make decisions about virtual warehouse sizing. Although a warehouse is simple to create, there are a few things to consider when deciding on the correct sizing. QUERY_HISTORY and WAREHOUSE_UTILIZATION show how a warehouse is running and provide metrics to assist in changing the configuration if required.

Table 5-2. Systems Metrics to Monitor

Metrics Name	Description
Bytes scanned	The more bytes scanned to process a query, the larger the warehouse needed. If most queries scan less than a GB, a small or extra small warehouse is likely sufficient.
Execution time	High execution time indicates long-running queries. Optimize these queries or scale up to a larger warehouse.
Queued time	A query competing with others for resources will have high queued times. Increase warehouse size or move queries to a less utilized warehouse.
Bytes spilled to local and bytes spilled to remove	This indicates that the warehouse is undersized. Bytes that can't fit into memory spill to disk storage, which is slower on local storage and even slower on remote storage.

(continued)

Table 5-2. (*continued*)

Metrics Name	Description
Out of memory	A query too large to fit in memory. Solutions include reducing concurrent queries, increasing maximum clusters, or using a larger warehouse.
Partition scanned and total	Partitions read by the query against the total number of micro-partitions in existence for the tables. Clustering and query join and predicate conditions help avoid reading unwanted data from disk.
Cache usage	Percentage of data scanned from the local warehouse disk cache. For near real-time applications, increase the auto-suspend time if the usage is low.

With all of this information, you can clearly identify workload patterns such as:

- What are the appropriate VW settings for a workload?
- Are there concurrency spikes in the workload?
- What is the range of query types? Is the size of the queries consistent or widely ranging?
- Are there peak loads on certain days of the month, quarter, and year?
- How often and how complex are ad hoc queries?
- Which user(s) perform which categories of queries?
- What types of queries run during business hours, but almost never on weekends or between 8 PM and 7 AM?

As you encounter best practices in warehouse sizing, attribute cost, and concurrency management, it is essential to apply these learnings to address the factors that affect warehouse utilization. The next section discusses finding the actual usage.

Warehouse Provisioned Capacity: Used vs. Wasted

To effectively manage warehouse capacity, it is crucial to understand how much of what you're paying for is actually being used. This insight allows you to identify and reduce waste while maximizing efficiency. In Snowflake, virtual warehouses can be provisioned for various purposes, such as isolating workloads, defining different scaling characteristics for development versus production environments, and even tracking credit consumption by application or department.

Snowflake charges for compute resources based on the total time that virtual warehouses are active in your account. This pricing model is simple, transparent, and easy to measure. However, there's an important consideration: the actual work being done, or the utilization, is not directly tied to the billing. For example, if you start a large warehouse and leave it idle for an hour, you'll be charged the same amount as if the warehouse were actively processing queries during that time.

This is where understanding and managing warehouse activity becomes important. For instance, in Figure 5-5, you can see queries represented by blocks. The blocks indicate the active user sessions that submitted the query and the concurrency level of the warehouse.

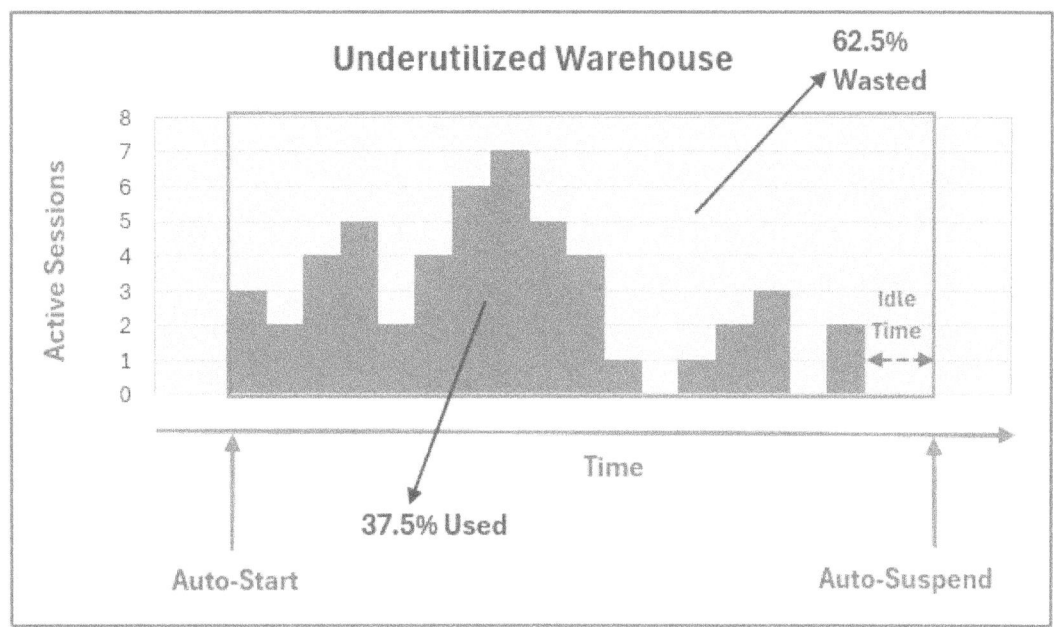

Figure 5-5. *Underutilized warehouse*

Here, the total area of the rectangle represents the capacity charged for, and the dark shaded area represents the capacity utilized. Snowflake warehouses can process eight concurrent queries by default, which sets the height of the rectangle. Note that the warehouse only briefly reached a concurrency of seven queries, with the rest of the time being much lower. The light area within the rectangle represents "wasted" capacity— capacity that was available (and paid for) but never used.

To make it clearer, if you could run all the queries from the previous scenario with maximum concurrency supported, you could save almost 50% of the compute for the same work. This is shown in Figure 5-6. Consolidating the workload is very important to reduce warehouse waste.

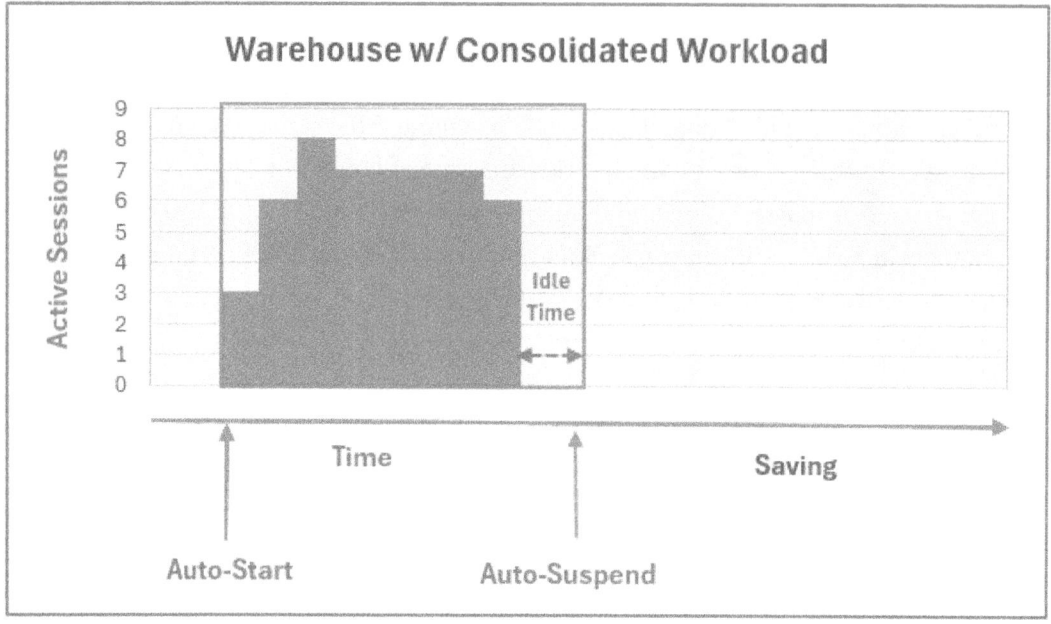

Figure 5-6. *Warehouse with consolidated workload*

However, when provisioning virtual warehouses for various purposes, such as isolating workloads, defining different scaling characteristics between development and production environments, and tracking credit consumption by application or department, it's important to consider efficiency. Configuring similar workloads to run on the same warehouse can help save compute resources and reduce costs.

Consolidate Workload

Warehouse utilization trends over a day or week can help identify opportunities to significantly reduce compute resourcec waste and improve warehouse utilization, as illustrated in Figure 5-7. Consider these key strategies for consolidating workloads:

> **Identify underutilized warehouses:** Analyze warehouse utilization trends to spot warehouses with low activity or outlier queries causing spikes in utilization.

CHAPTER 5 COMPUTE OPTIMIZATION STRATEGIES

Move outlier queries: Relocate resource-intensive queries to a dedicated warehouse sized appropriately for their needs. Downsizing the original warehouse can improve efficiency.

Combine similar workloads: Merge moderately utilized warehouses with similar workloads into a single multi-cluster warehouse for dynamic scaling or a standard warehouse for consistent workloads.

Provision common ad hoc warehouses: For teams frequently executing ad hoc queries across multiple warehouses, create a shared ad hoc warehouse to centralize queries and improve efficiency. If queuing increases during peak times, consider using a multi-cluster warehouse.

Optimize workload scheduling: Consolidate workloads based on usage patterns, such as business hours versus off hours. Use scheduled tasks to scale down warehouse size during non-business hours and reset it during peak times.

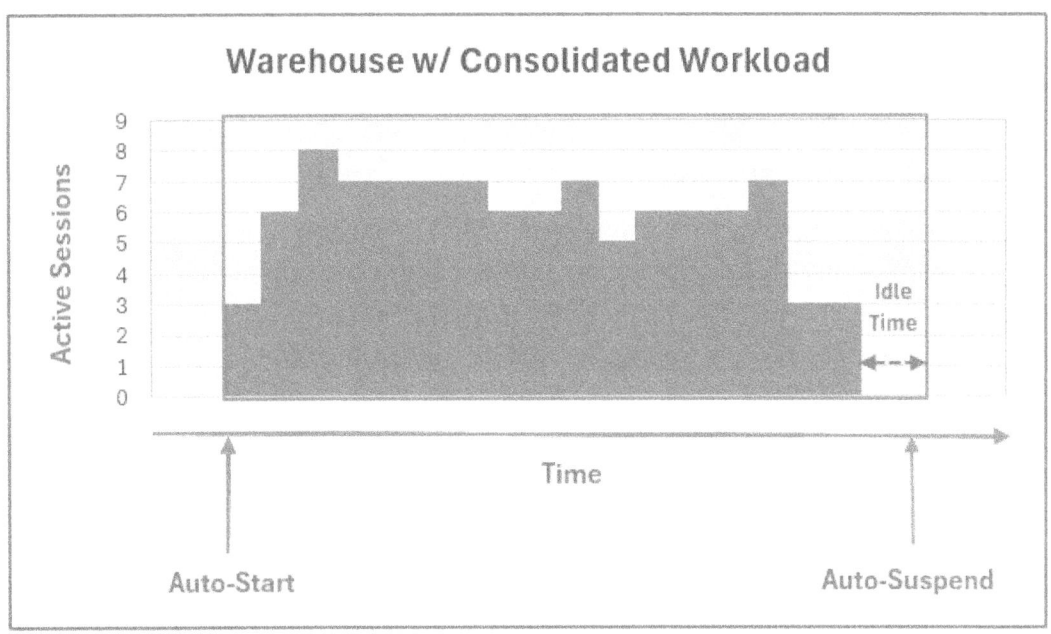

Figure 5-7. Optimally utilized warehouse

To optimize warehouse provisioning, it's not enough to track usage, you also need to understand how that usage occurs. The warehouse utilization heatmap helps reveal inefficiencies and opportunities for improvement.

Warehouse Utilization Heatmap

Understanding how efficiently your Snowflake warehouse is being used is crucial for optimizing costs and performance. While Snowflake doesn't directly show metrics like CPU, I/O, or memory usage for active warehouse clusters, you can uncover inefficiencies by analyzing views such as QUERY_HISTORY, WAREHOUSE_UTILIZATION, and QUERY_ATTRIBUTION_HISTORY.

Here is how you can approach this:

> **Analyze utilization metrics:** Use views like QUERY_HISTORY and WAREHOUSE_UTILIZATION to gain insights into key factors such as CPU usage, network I/O, memory consumption, and concurrent sessions. These metrics reveal patterns that can help you pinpoint inefficiencies.
>
> **Spot underutilization:** If your warehouse is consistently operating below 60% utilization, it's likely oversized. Resizing it to a smaller configuration can save on costs while maintaining adequate performance.
>
> **Visualize activity with a heatmap:** Build a heatmap using utilization data to visualize warehouse activity across different times of the day. This can help you identify peak usage periods and times when the warehouse is idle or underutilized.
>
> **Act on usage trends:** For warehouses with high activity during business hours but low usage after hours, consider consolidating workloads or setting up scheduled tasks to scale down the warehouse size during off hours. This ensures that resources are allocated efficiently without unnecessary costs.

Any indication of high utilization from the following query can indicate that the warehouse is overloaded, which could lead to queuing:

CHAPTER 5 COMPUTE OPTIMIZATION STRATEGIES

```
--Warehouse Utilization
WITH cte_wh_util AS
  (   SELECT *
      FROM
      (SELECT warehouse_name,
             HOUR(start_time) AS hour_start_time,
             IFNULL(AVG(utilization),0) AS avg_utilization
        FROM SNOWFLAKE.ACCOUNT_USAGE.WAREHOUSE_UTILIZATION
        WHERE start_time >= DATEADD(day, -1, CURRENT_TIMESTAMP)
        GROUP BY ALL) A
        PIVOT(SUM(avg_utilization) FOR hour_start_time IN (ANY ORDER BY hour_start_time))
  )
  SELECT l.*,
      SUM(a.credits_used_compute) AS total_credits
    FROM SNOWFLAKE.ACCOUNT_USAGE.WAREHOUSE_METERING_HISTORY a
    JOIN cte_wh_util l
      ON a.warehouse_name = l.warehouse_name
    WHERE a.start_time::date = DATEADD(day,-1,CURRENT_DATE)
    GROUP BY ALL
ORDER BY total_credits DESC;
```

As shown in Figure 5-8, analyzing heatmap trends allows you to right-size your warehouses, reduce waste, and enhance performance. By identifying periods of underutilization or overload, you can make informed decisions about scaling, scheduling, and workload distribution. Remember, every credit saved contributes to a more cost-efficient and responsive Snowflake environment.

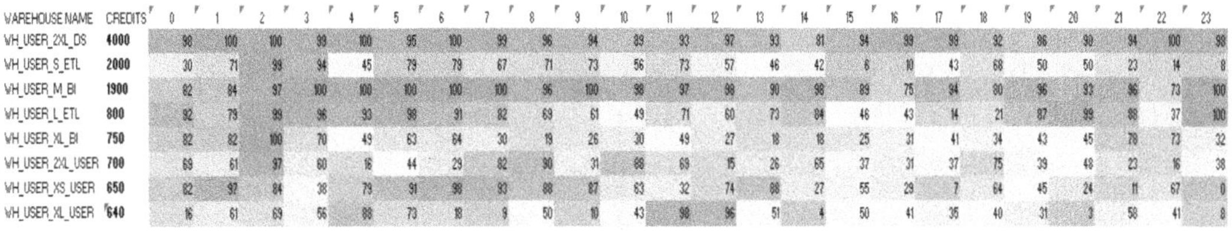

Figure 5-8. *Usage heatmap*

While heatmaps help visualize when and how warehouse resources are used, they don't reveal which queries are driving that usage. To gain deeper insight into workload efficiency and cost drivers, you can use the Query Attribute Cost metric. This metric allows you to attribute compute costs directly to individual queries, enabling more precise analysis and optimization.

Query Attribute Cost

Attribute cost serves as a crucial metric for analyzing shifts in workload over time, particularly when compute capacity is over-provisioned, making these changes difficult to detect. Based on our experience, the most effective strategy involves daily reporting on warehouse efficiency, complemented by detailed analytics on attribute cost. This combination of metrics enables you to optimize provisioning, minimize waste, and gain a comprehensive understanding of the key factors driving Snowflake usage.

The per-query cost attribution feature sets down the portion of the warehouse cost attributable to a specific query. When a single query is running in a warehouse, the entire warehouse cost is assigned to that query. Conversely, if multiple queries are running simultaneously, the warehouse cost during overlapping periods is distributed among the queries based on their relative resource consumption. During periods when the warehouse is idle (not suspended) and no queries are running, costs are not attributed to any query but can be easily determined and allocated across queries as needed for reconciliation purposes.

By leveraging attribute cost, you provide the application team with visibility into recurring, runaway, failure, and long-running queries. This insight allows you to address wasted resources and facilitate workload consolidation promptly.

To determine the attribute cost at the user level, utilize the following query (the following image shows an example):

```
WITH wh_bill AS (
   SELECT SUM(credits_used_compute) AS compute_credits
   FROM snowflake.account_usage.warehouse_metering_history
   WHERE start_time >= DATE_TRUNC('MONTH', DATEADD(MONTH, -1, CURRENT_DATE))
   AND start_time < DATE_TRUNC('MONTH', CURRENT_DATE)
),
```

```
user_credits AS (
    SELECT user_name, SUM(credits_attributed_compute) AS credits
    FROM snowflake.account_usage.query_attribution_history
    WHERE start_time >= DATE_TRUNC('MONTH', DATEADD(MONTH, -1,
CURRENT_DATE))
    AND start_time < DATE_TRUNC('MONTH', CURRENT_DATE)
    GROUP BY user_name
),
total_credit AS (
    SELECT SUM(credits) AS sum_all_credits
    FROM user_credits
)
SELECT u.user_name, u.credits / t.sum_all_credits * w.compute_credits AS
attributed_credits
FROM user_credits u, total_credit t, wh_bill w;
```

	USER_NAME	ATTRIBUTED_CREDITS
1	USER_MIKE	1.13
2	USER_JOHN	11.09
3	VELNATSF	29.39
4	USER_KING	45.11

Do Large Warehouses Cost More?

It is not always true that a larger warehouse size will result in higher credit costs when running queries. In fact, there are several factors that can make a larger warehouse size more cost-effective, depending on the nature of the queries.

How Query Composition Affects Warehouse Processing

The size of the warehouse influences how queries are processed, especially as the complexity and size of the query grow. In general, queries scale linearly with warehouse size, meaning the larger the warehouse, the more servers are available to handle the workload. This becomes particularly important when dealing with complex or large queries that require more computational power.

Impact on Query Performance

Larger warehouses can speed up query execution times, especially for more resource-intensive queries. Since Snowflake allocates more compute resources as the warehouse size increases, the performance of complex queries tends to scale linearly with the warehouse size. As a result, larger warehouses can typically process queries faster.

Now, here's the key takeaway: increasing the size of the warehouse doesn't necessarily increase the overall compute cost for a query. In fact, the total cost can remain roughly the same, as the increase in credits consumed by running the query is often balanced by the reduction in execution time. This means that scaling up the warehouse can lead to faster results without significantly increasing credit consumption.

For example, if you increase the warehouse size to handle a complex query faster, the additional time saved in query execution often outweighs the small overhead of suspending or resuming the warehouse. In many cases, this approach can be more cost-efficient than running the same query on a smaller warehouse for a longer period.

Overall, as illustrated in Figure 5-9, while it is easy to assume that larger warehouses automatically lead to higher costs, they can actually be a smart choice for improving query performance without significantly increasing credit consumption. The key is understanding how warehouse size aligns with the complexity and demands of the queries.

Complex queries	Small	Medium
Query Time	2 hrs	1 hr
Credits	2 x 2 / hr = 4	1 x 4 / hr = 4

Figure 5-9. Warehouse size over query performance

Factors that Affect Warehouse Utilization

Effective warehouse utilization in Snowflake is critical for optimizing performance, reducing costs, and ensuring that data operations are both efficient and scalable. Several factors can significantly impact the resource utilization of a Snowflake data warehouse. Understanding and managing these factors is essential for maintaining optimal warehouse performance, maximizing cost-efficiency, and delivering faster insights.

Snowflake's data warehousing solution provides flexibility, scalability, and performance, but several factors influence how effectively warehouse resources are utilized. Understanding these factors can help organizations optimize their operations and avoid inefficiencies. Key contributors to monitor in Snowflake warehouse utilization are shown in Figure 5-10.

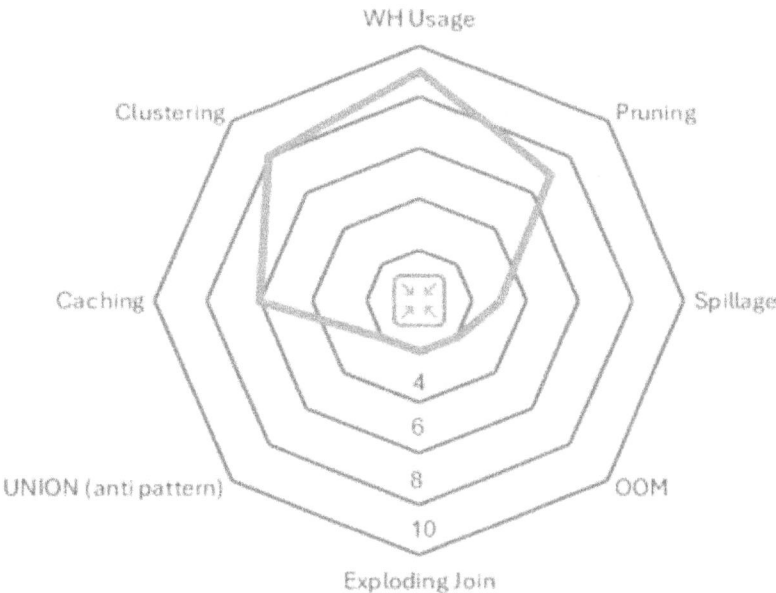

Figure 5-10. Key metrics to monitor on warehouse utilization

Row Explosion: How to Prevent Query Performance Issues

One of the most frequent causes of slow queries is a phenomenon known as "row explosion." This occurs when a join operation generates more rows than anticipated, leading to unwieldy datasets that degrade performance and waste processing power.

CHAPTER 5   COMPUTE OPTIMIZATION STRATEGIES

A row explosion typically happens when two tables are joined in such a way that every row from one table is combined with every row from the other table, often due to a missing or incorrect join condition. For example, in equality joins (such as 1:1 or 1:many relationships), the number of rows in the result should not exceed the size of the largest table involved. However, outer joins and many-to-many relationships can inflate the result set's cardinality beyond the size of either of the original tables, indicating a row explosion.

To prevent these issues, it is crucial to ensure that joins are properly written. Verify that the join keys are correct and accurately represent the relationships between the tables. Defining the join condition correctly on the business keys will help avoid the unexpected multiplication of rows, known as row explosion.

As shown in Figure 5-11, a simple way to identify row explosion issues is by examining the query profile. Look for join operators that show a higher number of output rows than the input tables. This usually occurs due to a high number of duplicates in the columns in the join condition. An extreme case is when the join condition is missing entirely.

When you identify a row explosion, consider the following:

- Why the join explodes

- Whether it is causing a performance concern (the larger the value of `row_multiple`, the more likely it is to lead to a problem)

- Whether the join explosion is required for the semantics of the query, or if it can be removed with a rewrite

- Whether the join explosion is "encoded" in the query, or if it has been introduced deliberately by a join-order bug in a compiler

There are instances when this is the intended behavior of the query author. However, in most cases, it results from an incorrect join condition.

Best scenario:

```
SELECT *
FROM DEMO_DB.TPCH_SF100.orders AS o
    JOIN DEMO_DB.TPCH_SF100.customer AS c
    ON o.o_custkey=c.c_custkey
WHERE c_custkey IN (121361, 84973, 47423)
;
```

Row explosion scenario:

```
SELECT *
FROM DEMO_DB.TPCH_SF100.orders AS o
    JOIN DEMO_DB.TPCH_SF100.customer AS c
WHERE c_custkey in (121361, 84973, 47423)
;
```

Figure 5-11. Row explosion join

This query will return a result for every join with a row_multiple over 1.0 that represents join explosions, and should be examined further (the following image shows an example).

CHAPTER 5   COMPUTE OPTIMIZATION STRATEGIES

```
--Programmatic Access to Query Profile
SET qid='01bc44ee-0002-d69e-0000-000813e634f5';
SELECT
   QUERY_ID, STEP_ID,OPERATOR_ID,
   OPERATOR_ATTRIBUTES:table_name::string tablename,
   OPERATOR_STATISTICS:output_rows output_rows,
   OPERATOR_STATISTICS:input_rows input_rows,
   CASE WHEN operator_statistics:input_rows>0
        THEN operator_statistics:output_rows / operator_statistics:input_rows ELSE 0 END
   AS JOIN_EXPLOSION_RATE
FROM table(get_query_operator_stats($qid))
HAVING JOIN_EXPLOSION_RATE > 1
ORDER BY STEP_ID,OPERATOR_ID;
```

QUERY_ID	STEP_ID	OPERATOR_ID	TABLENAME	OUTPUT_ROWS	INPUT_ROWS	JOIN_EXPLOSION_RATE
01bc44ee-0002-d69e-0000-000813e6	1	1	null	450000000	150000003	2.99999994

Strategies for Efficient Resolution

To improve performance in this scenario, consider implementing one or more of the following strategies:

- Verify join conditions carefully, ensuring business keys are used correctly to avoid unintentional many-to-many joins or cross joins.

- Use the query profile to identify joins with disproportionately high output rows compared to input. These are signs of row explosion and possible join logic issues.

- Rewrite or refactor queries to reduce unnecessary joins, deduplicate join columns, or restructure the query logic when row multiplication isn't essential to the business logic.

Disk Spilling (Paging)

Disk spilling happens when queries are too large to fit in memory.

When a Snowflake warehouse cannot fit an operation in memory, it starts spilling data, first to the local disk of a warehouse, and then to remote storage. In such a case, Snowflake first tries to temporarily store the data on the warehouse local disk. As this means extra IO operations, any query that requires spilling will take longer than a similar query running on similar data that is capable of fitting the operations in memory. Also, if the local disk is not big enough to fit the spilled data, Snowflake further tries to write to the remote cloud storage, which will be shown in the query profile as "Bytes spilled to remote storage".

Performance degrades drastically when a warehouse runs out of memory while executing a query because memory bytes must "spill" onto local disk storage. If the query requires even more memory, it spills onto remote cloud-provider storage, which results in even worse performance.

When the workload results in a lot of data scanned (reflected in the bytes_scanned column) or in high spillage to remote or local storage, the warehouse might be spending a lot of time and resources reading and writing data.

Run the following query to detect local and remote disk spills triggered by excessive memory usage (the following image shows an example).

```sql
SELECT query_id, user_name,
       warehouse_name,
       bytes_spilled_to_local_storage,
       bytes_spilled_to_remote_storage,
       query_text
FROM   snowflake.account_usage.query_history
WHERE  (bytes_spilled_to_local_storage > 0
  OR   bytes_spilled_to_remote_storage > 0 )
  AND  start_time >= DATEADD(day, -7, CURRENT_TIMESTAMP())
ORDER BY bytes_spilled_to_remote_storage, bytes_spilled_to_local_storage DESC
LIMIT 100;
```

USER_NAI	WAREHOUSE_NAME	BYTES_SPILLED_TO_LOCAL_STORAGE	BYTES_SPILLED_TO_REMOTE_STORAGE
VELNATSF	TPCDS_BENCH_10T	427687288832	0
VELNATSF	TPCDS_BENCH_10T	283163127808	0
VELNATSF	TPCDS_BENCH_10T	21953003520	0

Strategies for Efficient Resolution

To improve performance in this scenario, consider implementing one or more of the following strategies:

- Optimize queries to reduce memory load by limiting selected columns, improving partition pruning, and breaking complex logic into smaller steps using temporary tables.

- Control parallelism and workload concurrency to prevent excessive memory pressure that triggers spilling.

- Scale warehouse size or enable the QAS to provide more memory and disk capacity, minimizing both local and remote disk spilling.

- If utilization is low, with no data spilling and minimal bytes_scanned, consider downsizing the warehouse or reducing cluster count (for multi-cluster setups) to improve cost-efficiency without compromising performance.

Out of Memory (OOM)

When a query consumes more memory than the local virtual warehouse can provide, it triggers spilling to a local or remote disk. The VW can hit *out of memory* (OOM) when it does not respond quick enough to the local/remote disk. Snowflake often tries to rerun the query and occasionally the query fails after a few attempts. Expensive memory operations can include workload/query concurrency, large amounts of data processing, hash join builds, aggregation (grouping sets), window functions, and recurring CTEs.

Run the following query to detect local and remote disk spills triggered by excessive memory usage (the following image shows an example).

```
--Find Queries run into OOM
SELECT USER_NAME,START_TIME::DATE AS THE_DT,EXECUTION_TIME,
       QUERY_RETRY_TIME, QUERY_RETRY_CAUSE, FAULT_HANDLING_TIME
FROM SNOWFLAKE.ACCOUNT_USAGE.QUERY_HISTORY
WHERE START_TIME::DATE >=DATEADD('DAYS',-7,CURRENT_DATE())
AND QUERY_RETRY_TIME > 1
ORDER BY QUERY_RETRY_TIME DESC;
```

Row	USERNAME	THE_DT	EXECUTION_TIME	QUERY_RETRY_TIME	QUERY_RETRY_CAUSE
1	HLMQQ1W	2024-05-28	6492967	12472217	Warehouse Out Of Memory Error: Possible solutions include re...
2	SG3HKJH	2024-05-28	3412377	7312730	Warehouse Out Of Memory Error: Possible solutions include re...
3	BBVBVHW	2024-05-27	6010446	4775009	Warehouse Out Of Memory Error: Possible solutions include re...
4	JICLYNA	2024-05-29	1969797	2355385	Warehouse Out Of Memory Error: Possible solutions include re...
5	RTGTUEI	2024-05-29	1611460	2278234	Warehouse Out Of Memory Error: Possible solutions include re...
6	THDAPLI	2024-05-30	1498286	1987519	Warehouse Out Of Memory Error: Possible solutions include re...
7	XCUCZZ7	2024-05-27	426562	1248828	Warehouse Out Of Memory Error: Possible solutions include re...
8	PTWRXHO	2024-05-28	379664	998244	Warehouse Out Of Memory Error: Possible solutions include re...
9	STL8MAM	2024-05-30	14126	881292	Warehouse Out Of Memory Error: Possible solutions include re...

Strategies for Efficient Resolution

To improve performance in this scenario, consider implementing one or more of the following strategies:

- Split complex queries by materializing large intermediate results into temporary tables to reduce memory pressure during joins, aggregations, or window functions.
- Optimize pruning and expressions by simplifying filters and leveraging effective clustering to minimize unnecessary data scans.
- Reduce concurrency or scale up the warehouse size as a last resort to provide more memory per query and avoid retry/failure scenarios.

High Scanners (Using Partitions Scanned as a Heuristic)

High scanner queries are those that scan a large number of micro-partitions (MPs), often leading to long execution times and inefficient resource usage. A useful heuristic to detect such queries is examining the "partitions scanned" in the query profile. When most or all of the available partitions are scanned, it indicates that pruning is ineffective, often due to poor filter design, lack of clustering, or unselective query conditions.

Run the following query to detect high scan ratio queries, often resulting from missing filters or scanning large tables the following image shows an example).

```
SELECT
 query_id,
 user_name,
```

```
partitions_scanned,
partitions_total,
(partitions_scanned / partitions_total) as scan_ration,
query_text
FROM snowflake.account_usage.query_history
WHERE partitions_total > 0
AND (partitions_scanned / partitions_total) > 0.8;
```

USER_NAM	PARTITIONS_SCANNED	PARTITIONS_TOTAL	SCAN_RATION
VELNATSF	145436	72722	1.999890
VELNATSF	72721	72721	1.000000
VELNATSF	72721	72721	1.000000

Strategies for Efficient Resolution

To improve performance in this scenario, consider implementing one or more of the following strategies:

- Improve filter selectivity and pushdown by rewriting query filters (e.g., WHERE conditions) to be simpler and more selective, ensuring they align with clustering keys to enhance pruning efficiency.

- Recluster tables on frequently filtered columns. If the query regularly filters on specific columns (e.g., INVOICE_DATE, ITEM_KEY), define clustering on those columns to minimize the number of micro-partitions scanned.

- Leverage partition pruning diagnostics by using the query profile to monitor "partitions scanned" versus "total partitions" to assess pruning effectiveness and identify queries that behave as high scanners.

Warehouse Caching

When a Snowflake warehouse is active, it keeps a cache of frequently accessed table data to speed up query processing. This allows subsequent queries to pull data from the cache rather than retrieving it from the cloud storage again. The size of the cache depends on the compute resources allocated to the warehouse. A larger warehouse has more compute resources and, therefore, a larger cache.

However, when a warehouse is suspended, this cache is cleared. As a result, any queries that run right after the warehouse resumes might experience slower performance since they need to fetch data from the tables instead of the cache. Over time, as the warehouse processes more queries, the cache will rebuild, and queries that benefit from it will perform better.

When deciding whether to suspend a warehouse or leave it running, there's a tradeoff to consider. Suspending the warehouse saves credits, but it also results in losing the cache, which could affect performance. You must balance the importance of performance for the workloads against the potential cost savings.

Cost Considerations

It's important to remember that a running warehouse consumes credits even if it's idle and not processing queries. Make sure the auto-suspend setting aligns with the system workload. For example, if a warehouse runs a query every 30 minutes, setting the auto-suspend to 10 minutes would be inefficient. The warehouse would be suspended before the next query runs, wasting credits while sitting idle without benefiting from a cached state.

Find Data Scanned from Cache

This information can help identify if certain queries, especially those that are frequent or similar, could benefit from improved cache usage. Cache-usage ratio helps determine whether certain queries, particularly those that are run frequently or are similar in nature, could benefit from better utilization of the cache.

> **Note** When the percentage of data scanned from the cache is low, optimizing the cache by adjusting the warehouse size can result in a performance boost.

Use the following query to check the percentage of data scanned from cache across all queries, broken down by warehouse. This represents the percentage of data scanned from the local disk cache. The value ranges from 0.0 to 1.0, so multiplying by 100 will give you the percentage (the following image shows an example).

```sql
SELECT warehouse_name
  ,COUNT(*) AS query_count
  ,SUM(bytes_scanned) AS bytes_scanned
  ,SUM(bytes_scanned*percentage_scanned_from_cache) AS bytes_scanned_from_cache
  ,SUM(bytes_scanned*percentage_scanned_from_cache) / SUM(bytes_scanned) AS cache_hit_ratio
FROM snowflake.account_usage.query_history
WHERE start_time >= DATEADD(day, -7, CURRENT_TIMESTAMP())
  AND bytes_scanned > 0
GROUP BY ALL;
```

WAREHOUSE_NAME	QUERY_COUNT	BYTES_SCANNED	BYTES_SCANNED_FROM_CACHE	CACHE_HIT_RATIO
COMPUTE_WH	27	667300312	214306264	0.3211541493
TPCDS_BENCH_10T	61	4106230932645	1579403238624	0.3846357559

Strategies for Efficient Resolution

To improve performance in this scenario, consider implementing one or more of the following strategies:

- Align auto-suspend settings with query frequency to avoid unnecessary cache loss and warehouse restart overhead.

- Right-size warehouses to improve cache capacity and hit ratio for frequent or similar queries, balancing performance gains against cost.

- Monitor cache usage using query history (`percentage_scanned_from_cache`) to identify opportunities where improved cache retention could enhance performance.

Auto-Clustering

When tables are created with clustering keys, automatic clustering can quickly lead to high credit consumption due to the continuous reclustering of data. This can become a concern, particularly when heavy initial data loads are involved. A more efficient strategy for handling large data imports is to use the INSERT-SELECT-ORDER-BY approach while temporarily suspending reclustering on the table. However, if users are unaware of this best practice, they may inadvertently rack up significant costs from reclustering operations.

To mitigate this, it's advisable to disable automatic clustering on tables as soon as they are created, especially in data labs or workspace schemas, where clustering might not yet be needed. Instead, enable clustering selectively in business schemas, where it's more appropriate for performance optimization.

Currently, there isn't a feature to suspend reclustering at the table, schema, or database level through parameters. While clustering can be turned off at the account or table level, account-level control is too broad, and table-level control requires the table to be created first. Schema- or database-level control isn't effective because some tables might need clustering while others don't.

To manage these nuances, set up an automated process to periodically detect clustered tables and manage their reclustering behavior. The following section explains a pseudocode approach for creating such a procedure or application.

Pseudocode for Managing Automatic Clustering

This procedure would run every X minutes to ensure that newly created tables are properly managed:

- Check for new tables created since the last execution and determine whether automatic clustering is enabled.

- Verify against a lookup table to identify which tables are allowed to have clustering enabled (based on prior justification or authorization).

- Identify an appropriate role for managing these tables and assign the necessary permissions to the procedure or application role.

- Adjust clustering settings by doing one of these:

  - Remove clustering keys if unnecessary.
  - Suspend reclustering if it's not justified for that table.

This approach helps prevent excessive credit consumption due to unnecessary reclustering and ensures that automatic clustering is applied where it truly benefits performance and cost efficiency.

To better understand the clustering activity within the account, it is useful to retrieve the automatic clustering history. This can help you assess how much clustering has occurred and whether it aligns with your expectations. For example, if you are concerned about unexpected clustering costs, reviewing the past week's activity can give you insights.

Use the following query to pull the clustering history for the past seven days (the following image shows an example):

```
SELECT *
  FROM TABLE(information_schema.automatic_clustering_history(
    date_range_start=>dateadd(D, -7, current_date),
    date_range_end=>current_date));
```

START_TIME	END_TIME	CREDITS_USED	NUM_BYTES_RECL	NUM_ROWS_RECLU	TABLE_NAME
2025-05-10 17:00:00.00(	2025-05-10 18:00:00.00(	0.000037500	0	0	SNOWFLAKE.LOCAL.AI_OBSERV/

Strategies for Efficient Resolution

To improve performance in this scenario, consider implementing one or more of the following strategies:

- Disable automatic clustering by default for new tables, especially in staging or data lab schemas, to avoid unnecessary reclustering costs during bulk data loads or experimentation.

- For large data loads, use INSERT ... SELECT ... ORDER BY to maintain clustering order manually and minimize post-load reclustering overhead.

CHAPTER 5 COMPUTE OPTIMIZATION STRATEGIES

- Implement a monitoring procedure to periodically audit newly created tables, cross-reference with an *allow list*, and disable or adjust clustering settings based on operational justification.

UNION Without ALL

While UNION ALL just combines the inputs without any filtering, UNION does the same thing but with the added step of eliminating duplicates.

- UNION ALL simply concatenates inputs
- UNION concatenates inputs and performs duplicate elimination

Use the following query to find queries that use UNION, but would benefit from using UNION ALL (the following image shows an example).

```
SET qid='01bc4502-0002-d69e-0000-000813e63875';

SELECT  STEP_ID, OPERATOR_ID,
        OPERATOR_TYPE,
        CASE WHEN OPERATOR_TYPE    = 'UnionAll'
            AND lag(OPERATOR_TYPE) OVER (ORDER BY OPERATOR_ID) = 'Aggregate'
THEN 1 ELSE 0 END AS UNION_WITHOUT_ALL
FROM table(get_query_operator_stats($qid))
ORDER BY STEP_ID,OPERATOR_ID;
```

# STEP_ID	# OPERATOR_ID	A OPERATOR_TYPE	# UNION_WITHOUT_ALL
1	0	Result	0
1	1	Aggregate	0
1	2	UnionAll	1
1	3	Filter	0

151

Strategies for Efficient Resolution

To improve performance in this scenario, consider implementing one or more of the following strategies:

- Choose UNION ALL over UNION when duplicate elimination is not required. This prevents the expensive de-duplication step and improves performance.

- Analyze query plans or use get_query_operator_stats to identify unwanted UNION usage and confirm whether an Aggregate node follows the UNIONALL, indicating duplicate removal.

- Refactor queries using UNION to UNION ALL where business logic permits, to reduce compute cost and execution time.

Summary

Traditional RDBMS platforms typically emphasize throughput and performance, often leading to inefficiencies due to over-provisioning resources for peak loads. Snowflake, however, offers virtually unlimited compute and storage resources, making cost the primary factor for effective workload management. Optimizing compute resources in Snowflake requires proactive management and continuous refinement, often referred to as "refactoring" the workload distribution.

This chapter focused on fine-tuning various aspects of your warehouse setup to ensure efficient and effective use of resources. You begin by adjusting settings to match usage patterns, consolidating workloads, and monitoring key metrics to minimize inefficiencies. This process is enhanced by Snowflake's flexible architecture, which supports dynamic scaling, making it easier to manage costs while maintaining high performance.

In the next chapter, we explore deeper techniques to enhance efficiency. This includes analyzing rarely used tables, materialized views, search paths, and high-impact queries. Additionally, we delve into techniques for handling semi-structured data and external tables to further optimize your Snowflake environment.

CHAPTER 6

Workload Optimization Strategies

Overview

This chapter explores a comprehensive set of workload optimization strategies designed to help data teams maximize performance while maintaining control over compute and storage costs. From leveraging query patterns to mitigating query skew and optimizing the consumption layer, this chapter provides actionable techniques for tuning Snowflake workloads. It also compares ETL and ELT approaches, introduces Snowpark for in-platform data processing, and outlines best practices for consumption layer and intra-row calculations. Whether you manage batch pipelines or support real-time analytics, these strategies are essential for sustainable and scalable operations.

The strategies covered in this chapter include:

- Identifying high-frequency query patterns that silently drive up compute costs.

- Detecting and mitigating query and join skew to ensure balanced compute distribution.

- Choosing between ETL and ELT architectures based on data volume, transformation complexity, and cost-efficiency.

- Leveraging Snowpark to reduce data movement and eliminate the need for external processing engines.

- Designing the consumption layer for scalability, performance, and cost control.

- Learning from real-world incidents such as unexpected credit surges to implement proactive monitoring and cost controls.

Before diving into specific optimization techniques, it is essential to understand which queries are silently driving up costs. The first step is identifying high-cost query patterns that may not be obvious at first glance. This foundational step sets the stage for improving system performance and achieving robust cost management.

Identifying High-Cost Query Patterns

Compute usage is directly tied to cost. While long-running queries are easy to spot, they are not always the most expensive. A short query that runs thousands of times a day can quietly become a major contributor to warehouse consumption. To uncover these hidden cost drivers, Snowflake provides two query patterns to identify such high cost patterns.

Analyzing Query Patterns with Query Hash

The query_hash value is generated based on the exact SQL text of a query. As shown in Table 6-1, if a query is submitted multiple times without any changes to its structure or values, it will have the same query_hash. This makes it easy to group and analyze identical queries over time.

Table 6-1. Pattern Matching Query Hash

Query Text	query_id	query_hash
SELECT * FROM sales_data LIMIT 10;	q_001	hash_abc
SELECT * FROM sales_data LIMIT 10;	q_002	hash_abc
SELECT * FROM sales_data LIMIT 10;	q_003	hash_abc

These metrics help you track queries that are exactly the same in structure and content. Each execution has a unique query_id, but the query_hash is identical.

CHAPTER 6 WORKLOAD OPTIMIZATION STRATEGIES

This allows Snowflake to group these executions together and analyze their collective impact on performance and cost.

Use cases for query_hash:

- Track how often a specific query is executed.

- Measure total and average execution time for repeated queries.

- Compare performance before and after optimization efforts.

- Identify queries that are frequently repeated and consuming significant compute time.

This approach is especially useful for dashboards, scheduled jobs, or applications that issue the same query repeatedly.

The following query helps analyze queries that follow a common pattern (the following image shows an example).

```
SELECT
  query_hash,
  COUNT(*) AS executions,
  (AVG(total_elapsed_time)/60000)::NUMBER(12,2) AS avg_minutes,
  MAX(QUERY_TEXT) AS QUERYTXT
FROM snowflake.account_usage.query_history
WHERE start_time >= CURRENT_DATE() - 7
GROUP BY ALL
ORDER BY executions DESC;
```

QUERY_HASH	EXECUTIONS	AVG_MINUTES	QUERYTXT
b904fe352a2f6191befa8e6f8f129745	32	0.50	SELECT * FROM TPCH_SF1.CUSTOMER WHERE C_NATIONKEY=2;
c77d05caf091028e3e9ac6bb84d15fe4	23	1.70	SELECT * FROM TPCH_SF1.CUSTOMER WHERE C_NATIONKEY=1;
59ec19edb8aa152c1167929ad63b7763	20	2.70	SELECT * FROM TPCH_SF1.CUSTOMER WHERE C_NATIONKEY=15;
9fe2ecb286d04213695895b823d29da2	18	0.80	SELECT * FROM TPCH_SF1.CUSTOMER WHERE C_NATIONKEY=10;
24f927b159bd8a674fbec85bab8b57d0	17	0.50	SELECT * FROM TPCH_SF1.CUSTOMER WHERE C_NATIONKEY=3;

Analyzing Query Patterns with Query Parameterized Hash

While query_hash groups identical queries, query_parameterized_hash goes a step further. As shown in Table 6-2, it groups queries that have the same structure but differ only in literal values, such as numbers or strings in the WHERE clause.

CHAPTER 6 WORKLOAD OPTIMIZATION STRATEGIES

Table 6-2. Parameterized Query Hash

Query Text	query_hash	query_parameterized_hash
SELECT * FROM sales WHERE region = 'US'	hash_a	paramhash_x
SELECT * FROM sales WHERE region = 'EU'	hash_b	paramhash_x
SELECT * FROM sales WHERE region = 'APAC'	hash_c	paramhash_x

These metrics help you:

- See which query patterns ran the most
- Understand how much total time they consumed
- Spot opportunities to optimize or consolidate frequently used queries

Use cases for query_parameterized_hash:

- Group structurally similar queries that differ only in literal values (e.g., different dates, regions, or user IDs).
- Analyze performance trends across a family of similar queries.
- Identify common query patterns generated by applications or reporting tools.
- Optimize query templates that are frequently used with varying parameters.
- Reduce noise in query performance analysis by focusing on structure rather than specific values.

The following query helps analyze queries that follow a common pattern but vary slightly in input (the following image shows an example).

```
SELECT
  query_parameterized_hash,
  COUNT(*) AS executions,
  (AVG(total_elapsed_time)/60000)::NUMBER(12,2) AS avg_minutes,
  MAX(QUERY_TEXT) AS QUERYTXT
FROM snowflake.account_usage.query_history
```

```
WHERE start_time >= CURRENT_DATE() - 7
GROUP BY ALL
ORDER BY executions DESC;
```

QUERY_PARAMETERIZED_HASH	EXECUTIONS	AVG_MINUTES	QUERYTXT
c6903cf2c16b387482a2755539081aa9	110	1.80	SELECT * FROM TPCH_SF1.CUSTOMER WHERE C_NATIONKEY=3;
337620eecd6d2c27e17ef3757ff625df	33	0.50	SELECT * FROM TPCH_SF1.PART WHERE P_PARTKEY=40003;
daffb8c9d0733cc13b1ffb7297f863ce	15	0.60	SELECT * FROM TPCH_SF1.ORDERS WHERE O_ORDERSTATUS='F';
7c771f12db9a3a558afc14b72fafca8a	13	1.10	SELECT * FROM TPCH_SF1.ORDERS WHERE O_CUSTKEY=1481;

In this case, although the `nationkey` and `partkey` values differ, the structure of the query is the same. Snowflake recognizes this and assigns the same `query_parameterized_hash`.

Why Query Patterns Matter

Focusing only on slow queries can lead to missed opportunities for optimization. Query patterns help you:

- Spot high-frequency queries that may be short but costly in aggregate.
- Group similar queries to understand their collective impact.
- Track performance trends for specific query types over time.
- Prioritize tuning efforts based on actual usage patterns, not just duration.

By analyzing `query_hash` and `query_parameterized_hash`, you can better understand how workloads behave and identify areas where improvements can be made. After pinpointing expensive queries, the next step is to examine how data objects are being used. Tracking object usage helps manage computing and storage costs more efficiently.

Tracking Object Usage for Cost Management

Snowflake provides a powerful metadata view called ACCESS_HISTORY that enables you to track how database objects, such as tables, views, and schemas, are being accessed. This is particularly valuable for identifying underutilized or unused objects, optimizing query performance, and ultimately reducing compute and storage costs.

Why ACCESS_HISTORY Matters

Understanding which objects are frequently accessed and which are not can help data teams:

- Identify unused tables or views that can be archived or dropped.
- Spot high-usage objects that may benefit from clustering or materialization.
- Track access patterns to align compute resources with actual demand.
- Audit object usage for compliance and governance.

Find Table and View Usage

The following SQL query identifies table usage based on the number of queries in a specific database and schema (the following image shows an example).

```
with queries as
    (
    select
      query_id as query_id_qry_hst
      ,user_name
      ,role_name
      ,warehouse_name
      ,session_id
    from snowflake.account_usage.query_history
    where start_time >= dateadd('day',-30,current_date)
    order by session_id
```

```sql
    ),
accessLogs as
    (
      select
        query_id as query_id_acc_hst
        ,direct_objects_accessed
        ,doa.value:"objectName"::string as object_name
        ,doa.value:"objectDomain"::string as object_type
        ,split_part(object_name,'.',1) as database_name
        ,split_part(object_name,'.',2) as schema_name
        ,split_part(object_name,'.',3) as table_name
      from snowflake.account_usage.access_history,
      lateral flatten(direct_objects_accessed) doa
      where query_start_time >= dateadd('day',-30,current_date)
    ),
sessions as
    (
      select
        session_id
        ,parse_json(client_environment):APPLICATION::string as application
      from snowflake.account_usage.sessions
      where created_on::date >= dateadd('day',-30,current_date)
    )
select
    user_name
    ,role_name
    ,application
    ,object_name
    ,object_type
    ,database_name
    ,schema_name
    ,table_name
    ,count(distinct query_id_qry_hst) as query_count
```

```
from
    (select *
    from queries
    join accessLogs on queries.query_id_qry_hst = accessLogs.query_
id_acc_hst
    join sessions on queries.session_id = sessions.session_id)
where
    user_name = 'BI_LOOKER'
group by ALL
order by query_count desc;
```

OBJECT_NAME	OBJECT_TYPE	DATABASE_NAME	SCHEMA_NAME	TABLE_NAME	QUERY_COUNT
DEMO_DB.TPCH_SF1.CUSTOMER	Table	DEMO_DB	TPCH_SF1	CUSTOMER	111
DEMO_DB.TPCH_SF1.ORDERS	Table	DEMO_DB	TPCH_SF1	ORDERS	38
DEMO_DB.TPCH_SF1.PART	Table	DEMO_DB	TPCH_SF1	PART	34

For high-usage tables, explore clustering keys or materialized views to improve performance and reduce compute time.

After understanding how data objects are accessed, it's equally important to ensure that compute resources are used efficiently. Managing operational skew helps maintain balanced workloads and prevents performance bottlenecks across Snowflake's compute nodes.

Manage Operational Skew

In a massively parallel processing (MPP) system like Snowflake, performance and efficiency hinge on the even distribution of work across compute nodes within the warehouse. When this balance is disrupted, *operational skew* occurs, leading to inefficient resource utilization, slower query performance, and potential contention with other concurrent workloads.

Operational skew arises when certain operations, such as filtering, joining, or grouping, are performed on columns with low cardinality. These are columns that contain only a few distinct values, often with one value (such as NULL) dominating the dataset. In such cases, Snowflake's query planner may assign disproportionate amounts of data to a subset of compute nodes, causing those nodes to become bottlenecks while others remain underutilized.

Scenario: Skewed Column

Consider a `prescriptions` table where 40% of the rows have NULL in the `drug_id` column. If this column is used in a join or group-by operation, the compute node responsible for processing the NULL values will receive a significantly larger portion of the data. This leads to:

- Uneven workload distribution across compute nodes within the virtual warehouse.

- Increased query latency due to straggling compute nodes.

- Resource contention that can degrade the performance of other concurrent queries.

System-Wide Impact

In a shared compute environment, operational skew doesn't just affect the query in question. It can ripple across the system and lead to other problems:

- **Concurrency degradation:** Skewed queries may monopolize CPU and memory resources, delaying other workloads.

- **Inefficient scaling:** Simply adding more compute resources may not resolve the issue if the skew remains unaddressed.

- **Cost inefficiency:** Longer-running queries consume more credits, increasing operational costs without delivering proportional value.

Best Practices for Managing Skew

To mitigate operational skew in Snowflake:

- Analyze data distribution before using columns in joins or aggregations.

- Avoid low-cardinality columns as join keys or group-by targets when possible.

CHAPTER 6   WORKLOAD OPTIMIZATION STRATEGIES

- Use clustering keys to improve data locality and reduce skew-related overhead.
- Leverage query profiling tools (like Query Profile in Snowsight) to identify skewed execution patterns.

By proactively managing operational skew, organizations can ensure more predictable performance, better concurrency, and more efficient use of Snowflake's elastic compute resources.

Find Skewed Join and Aggregate Columns

Profiling join and aggregate columns helps detect data skew by analyzing value distributions. Use the following query to analyze join cardinality on key columns. This helps assess join performance and guides in taking corrective actions.

```
SELECT
    drug_id,
    COUNT(*) AS total_prescriptions
FROM
    prescriptions
GROUP BY ALL
ORDER BY 2 DESC;
```

The following query compares row counts for a join key across two tables, highlighting potential imbalances that can impact performance:

```
WITH left_summary AS (
    SELECT
        left_table_join_key AS key_value,
        COUNT(*) AS cnt_left
    FROM my_left_table
    GROUP BY ALL
),
right_summary AS (
    SELECT
        right_table_join_key AS key_value,
        COUNT(*) AS cnt_right
```

```
    FROM my_right_table
    GROUP BY ALL
)
SELECT
    l.key_value,
    l.cnt_left,
    r.cnt_right,
    l.cnt_left * r.cnt_right AS join_count
FROM left_summary l
JOIN right_summary r ON l.key_value = r.key_value
WHERE l.cnt_left * r.cnt_right > 1
ORDER BY join_count DESC
LIMIT 100;
```

How to Mitigate this Skew

This section looks at how to identify and fix skewed workloads to maintain consistent performance and make better use of compute resources.

> **Filter out NULLs if they are not needed:** Removing NULLs from join or group-by operations helps avoid overloading a single compute node with disproportionate data volume.
>
> ```
> SELECT
>     drug_id,
>     COUNT(*) AS total_prescriptions
> FROM prescriptions
> WHERE drug_id IS NOT NULL
> GROUP BY ALL;
> ```
>
> **Use conditional aggregation to separate NULLs:** Breaking out NULL and non-NULL values allows for more balanced processing and clearer insight into data distribution.

```
SELECT
    COUNT(*) FILTER (WHERE drug_id IS NULL) AS null_drug_count,
    COUNT(*) FILTER (WHERE drug_id IS NOT NULL) AS non_null_
drug_count
FROM prescriptions;
```

While managing compute efficiency is critical, minimizing data movement is equally vital for performance and cost control. Snowpark offers a powerful solution by enabling in-platform processing directly within Snowflake.

Reducing Data Movement Using Snowpark

In a data-driven economy, organizations face mounting pressure to process vast volumes of data with speed, security, and cost-efficiency. Traditional approaches often rely on external processing engines like Apache Spark, which introduce complexity, latency, and operational overhead. Snowpark, Snowflake's developer framework, is changing that paradigm.

As illustrated in Figure 6-1, Snowpark enables developers and data engineers to build and execute data-intensive applications directly within the Snowflake Data Cloud, eliminating the need to move data across platforms. This reduction in data movement not only enhances security by minimizing exposure but also significantly reduces costs associated with data transfers and infrastructure sprawl. Organizations gain deeper visibility into their cloud spend while streamlining operations within a unified, governed environment.

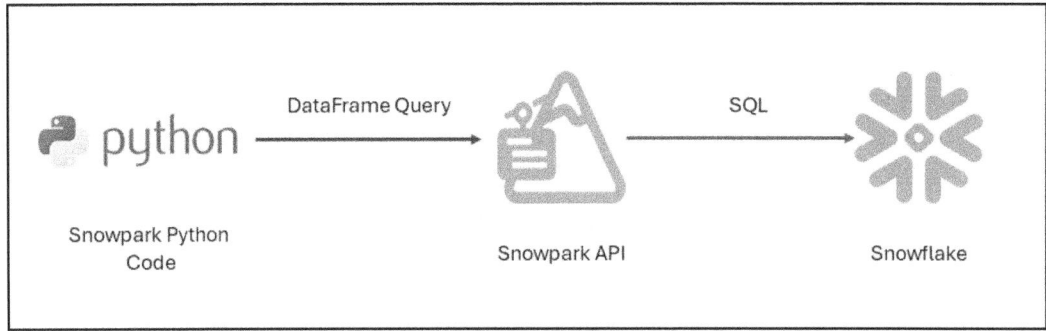

Figure 6-1. Snowflake Snowpark

CHAPTER 6   WORKLOAD OPTIMIZATION STRATEGIES

By executing all transformations natively within Snowflake's highly scalable compute layer, teams can process data faster and more predictably, even at enterprise scale.

A key enabler of this performance is the Snowpark-optimized compute node. These specialized virtual warehouses are designed to handle memory-intensive workloads and support advanced operations like machine learning training and user-defined functions (UDFs). With configurable memory and CPU architecture, Snowpark-optimized warehouses provide the flexibility and power needed for modern data applications, without the complexity of managing external clusters.

Finally, Snowpark simplifies the data ecosystem. Engineers no longer need to manage Spark clusters or orchestrate complex data pipelines across disparate systems. Snowpark offers several advantages over the traditional Snowflake Connector for Spark:

- **Language-native development:** Developers can use familiar tools like Jupyter, VS Code, or IntelliJ to write Snowpark code in Python, Java, or Scala.

- **Full pushdown support:** All operations, including UDFs, are pushed down to Snowflake, ensuring efficient execution at scale.

- **No external clusters:** All computation happens within Snowflake, eliminating the need for separate Spark clusters and simplifying infrastructure management.

To make the most of Snowpark and reduce data movement, Snowflake provides Snowpark-optimized warehouses designed for high-performance processing within the platform.

Snowpark-Optimized Warehouses

Snowpark-optimized warehouses are designed for advanced data processing and machine learning workloads. They offer significantly more memory and larger local cache compared to standard warehouses, making them ideal for Snowpark-based applications. These warehouses are tailored for compute- and memory-intensive tasks like training models, running UDFs, and handling large datasets efficiently; see Table 6-3.

Table 6-3. Snowpark-Optimized Warehouse Features

Feature	Standard Warehouse	Snowpark-Optimized Warehouse
Purpose	General-purpose SQL workloads	Optimized for Snowpark, ML, and memory-intensive tasks
Memory per node	Baseline (1x)	Up to 16 times more memory per node
Local cache	Standard	10 times larger local cache
CPU architecture	Default (varies by cloud provider)	Configurable: Default or x86
Minimum warehouse size	X-Small	Medium (256 GB), Large (1 TB) options
Performance	Suitable for most SQL workloads	Enhanced for ML training, UDFs, and large data processing
Elasticity and scaling	Yes	Yes
Startup time	Faster	Slightly longer startup time
Use case examples	BI dashboards, ETL, ad hoc queries	ML training, Python/Java UDFs, large data exports

As shown in Figure 6-2, these warehouses consume more credits per hour than standard virtual warehouses due to their increased resource allocation. This higher cost reflects the enhanced performance and memory they provide, making them ideal for workloads that demand greater computing power.

Snowpark-Optimized Warehouse Size (Credits / Hour)

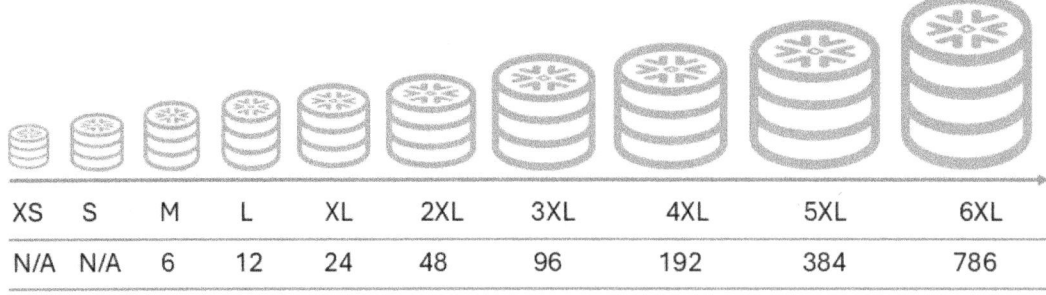

XS	S	M	L	XL	2XL	3XL	4XL	5XL	6XL
N/A	N/A	6	12	24	48	96	192	384	786

Figure 6-2. Snowpark optimized warehouse credits/hour

Example: Creating a Snowpark-Optimized Warehouse

```
CREATE OR REPLACE WAREHOUSE snowpark_opt_wh
  WITH WAREHOUSE_SIZE = 'LARGE'
  WAREHOUSE_TYPE = 'SNOWPARK-OPTIMIZED'
  ;
```

Snowpark Performance Best Practices

To maximize performance and minimize cost:

- **Minimize fields:** Select only necessary columns to reduce data movement.
- **Cache wisely:** Use `.cache()` only when transformations are reused.
- **Redistribute input for UDFs:** Use `df.sort(random())` to avoid skew and improve parallelism.

Example: Efficient DataFrame Usage

```
# Minimize fields
df = session.table("sales_data").select("region", "revenue")

# Cache only if reused
df_cached = df.cache()

# Redistribute data to avoid skew
from snowflake.snowpark.functions import rand
df = df.sort(rand())
```

As data volumes grow and transformation needs evolve, selecting the right pipeline architecture becomes critical. The next section explores how ETL and ELT approaches impact performance, scalability, and cost in Snowflake.

CHAPTER 6   WORKLOAD OPTIMIZATION STRATEGIES

Choosing Between ETL and ELT

Choosing between ETL (Extract, Transform, Load) and ELT (Extract, Load, Transform) is a critical decision when designing data pipelines in Snowflake. Each approach has distinct advantages depending on data volume, structure, transformation complexity, and infrastructure preferences. This section explores both strategies, helping you align your workload optimization with Snowflake's strengths in scalability, performance, and cost-efficiency.

ETL (Extract, Transform, Load)

Process: Extract ➤ Transform (outside Snowflake) ➤ Load into Snowflake

ETL is a traditional data integration approach where data is transformed before it enters Snowflake. This method is ideal for structured data and pipelines with strict transformation rules. It ensures that only clean, validated data reaches the warehouse, reducing the processing burden on Snowflake.

Best for: Structured data, strict transformation logic, pre-validated pipelines

Advantages:

- Offers full control over transformation logic using external tools.
- Minimizes compute load on Snowflake by offloading transformations.
- Ensures clean, preprocessed data is stored in the warehouse.

Challenges:

- Can be slower for large datasets due to external processing.
- Requires additional infrastructure and orchestration tools.
- Less adaptable for exploratory or ad hoc analysis.

ELT (Extract, Load, Transform)

Process: Extract ➤ Load raw data into Snowflake ➤ Transform using SQL

ELT leverages Snowflake's powerful compute engine to perform transformations after loading raw data. This approach is well-suited for high-volume or semi-structured data and simplifies pipeline architecture by reducing external dependencies.

Best for: High-volume, semi/unstructured data (e.g., JSON, logs)

Advantages:

- Utilizes Snowflake's MPP (massively parallel processing) engine for fast transformations.

- Enables rapid data ingestion and transformation within the same environment.

- Simplifies architecture by eliminating the need for external transformation tools.

Challenges:

- Storing raw data can increase storage costs.

- Requires strong SQL skills and governance to manage transformations effectively.

- Poorly optimized queries can lead to higher credit consumption.

Optimization Considerations

Choosing between ETL and ELT is not just a matter of architecture; it directly impacts performance, flexibility, and cost in Snowflake. Each strategy has tradeoffs that influence how compute resources are used, how data is transformed, and how efficiently workloads scale. Table 6-4 highlights the key considerations for selecting the right strategy when optimizing for both cost efficiency and performance.

Table 6-4. ETL vs. ELT Factors to Consider

Factor	ETL	ELT
Data volume	Low to medium	Medium to high
Transformation	Complex, external logic	SQL-based, scalable
Latency	Higher (sequential steps)	Lower (parallelizable)
Cost control	Lower compute, higher tool cost	Higher compute, lower tool cost
Flexibility	Rigid, predefined flows	Agile, supports raw data access
Tooling	Requires ETL tools (e.g., Ab Initio, Matillion)	Native SQL or dbt in Snowflake

Recommended Strategy

Use ETL when:

- You require strict control over data quality and transformation logic.
- External systems are already in place to handle heavy processing.
- Your workloads are predictable and structured, with minimal need for on-the-fly analysis.

Use ELT when:

- You want to maximize Snowflake's compute power for transformation.
- You're working with large, fast-moving, or semi-structured datasets.
- You need flexibility to iterate quickly and support ad hoc analytics.

Or consider a hybrid approach. Modern data architectures often blend both strategies. Use ETL for curated, compliance-heavy pipelines and ELT for high-volume, flexible ingestion. This hybrid model balances control, performance, and cost, allowing teams to optimize workloads based on specific business needs.

As pipeline complexity grows, so does the risk of performance bottlenecks. Breaking down complex transformations into modular steps can significantly improve query efficiency and maintainability in Snowflake.

Break Down Complex Transformations

In many data pipelines, it is common to see entire transformation workflows written as a single, deeply nested SQL query. While this may seem efficient, it can hinder Snowflake's ability to optimize performance.

Often, these queries reference views that are themselves built on top of other complex views. This layered design can obscure logic, making it difficult to maintain and even harder for Snowflake to accurately estimate the size of data involved in join operations, an essential factor in generating efficient query plans.

For example, Snowflake typically uses a hash join strategy, where the smaller dataset is loaded into memory and the larger one is streamed from storage. When multiple layers of joins and aggregations are involved, it becomes harder to determine which dataset is truly the largest. This can result in inefficient plans that consume excessive memory and processing time.

Recommended Strategy

Instead of executing all transformations in a single query, break the logic into smaller, modular steps using intermediate or transient tables. This approach offers several advantages:

- **Better performance:** Smaller queries are easier for the optimizer to handle.

- **Parallel execution:** Intermediate steps can run concurrently, reducing total runtime.

- **Improved maintainability:** Modular logic is easier to read, debug, and update.

- **Stable query plans:** Each step has a more predictable performance profile.

You can implement this using standard SQL or leverage Snowflake's Dynamic Tables to manage intermediate results declaratively. This modular approach not only improves performance but also enhances clarity and scalability, especially in batch processing scenarios where parallelism can significantly reduce execution time.

Example: Breaking Down a Transformation Pipeline

Instead of this monolithic query:

```
-- Complex and hard to optimize
SELECT c.customer_id, SUM(o.amount) AS total_spent
FROM customers c
JOIN orders o ON c.customer_id = o.customer_id
JOIN products p ON o.product_id = p.product_id
WHERE p.category = 'Electronics'
GROUP BY c.customer_id;
```

Break it into modular steps:

```sql
-- Step 1: Filter relevant products
CREATE OR REPLACE TEMP TABLE tmp_online_products AS
SELECT product_id
FROM sales_schm.products
WHERE category = 'Electronics';

-- Step 2: Join orders with filtered products
CREATE OR REPLACE TEMP TABLE tmp_online_orders AS
SELECT o.*
FROM sales_schm.orders o
JOIN tmp_online_products p ON o.product_id = p.product_id;

-- Step 3: Aggregate spend per customer
SELECT c.customer_id, SUM(o.amount) AS total_spent
FROM sales_schm.customers c
JOIN tmp_online_orders o
ON c.customer_id = o.customer_id
GROUP BY ALL;
```

This approach improves transparency, enables parallel execution, and gives Snowflake's optimizer better visibility into each step.

After optimizing transformations and pipeline logic, attention must shift to where most compute costs are incurred, the consumption layer. Designing this layer thoughtfully is the key to ensuring performance, scalability, and cost-efficiency in Snowflake.

Designing the Consumption Layer

Most costs will be incurred in the consumption layer. By an order of magnitude, these costs outweigh the costs of other layers, such as the ETL layer. Therefore, it is imperative to design this layer with goal of optimization and performance.

There are many factors to consider when designing the consumption layer, such as usability, maintainability, and deployment.

Focus on optimizing resource consumption, optimizing performance, and minimizing costs. Other considerations such as maintainability, ETL simplicity, and deployment agility, may weigh into the final decision as to whether to implement these recommendations.

Why the Consumption Layer Matters

Unlike the ETL or staging layers, which are typically batch-driven and predictable, the consumption layer supports ad hoc queries, dashboards, datamarts, and analytics workloads. These workloads are dynamic, user-driven, and often unpredictable, making this layer the most critical for performance tuning and cost control.

- **Cost impact**: By an order of magnitude, the consumption layer can generate significantly more credit usage than upstream layers.

- **User experience:** This is the layer most visible to analysts, data scientists, and business users. Performance here directly affects productivity.

- **Scalability pressure:** As more users and tools access Snowflake, the demand on this layer grows rapidly.

Key Design Principles

Optimize for performance and cost:

- Use materialized views or result caching for frequently accessed queries.

- Apply clustering keys on large tables to reduce scan costs.

- Leverage query acceleration services only when justified by performance needs.

Design for usability:

- Create semantic layers or curated datamarts that abstract complexity.

- Use consistent naming conventions and documentation to improve discoverability.

Ensure maintainability:

- Modularize SQL logic using views and stored procedures.
- Implement query tagging to track and analyze workload patterns.

Support deployment agility:

- Use version-controlled SQL scripts and CI/CD pipelines to deploy changes.
- Automate testing and validation of data models and transformations.

Monitor and iterate:

- Continuously monitor query performance using QUERY_HISTORY and WAREHOUSE_LOAD_HISTORY.
- Set up resource monitors to alert on unusual consumption patterns.

Intra-row calculations are a powerful tool in the Snowflake optimization toolkit, but like all optimizations, they must be applied judiciously to deliver meaningful performance and cost benefits.

Intra-Row Calculations

Computations performed within a single row of data can significantly impact query performance and resource consumption in Snowflake. When used strategically, precalculating and storing these values can reduce I/O, improve query speed, and lower compute costs. However, this optimization is most effective under specific conditions:

- **Precalculate when appropriate:** Store intra-row calculations if the table has over 1 million rows and the calculated fields are queried frequently, especially more often than the source columns.
- **Avoid overuse:** Adding too many precalculated fields can increase micro-partitions and I/O, especially on smaller warehouses, leading to diminishing performance returns.
- **Optimize string storage:** Store character data in a consistent format (e.g., uppercase and trimmed) to avoid runtime function overhead like UPPER() or TRIM().

CHAPTER 6 WORKLOAD OPTIMIZATION STRATEGIES

- **Performance benefits:** Precalculated fields can reduce I/O, improve query speed, and enable better partition pruning and optimizer decisions.

- **Evaluate tradeoffs:** While storage impact is minimal, precalculations require ETL changes and should be used only when they provide measurable performance gains.

Precalculation is recommended when queries operate on large datasets (typically over one million rows), and when those queries are executed frequently. If the calculated column is accessed more often than the raw columns used in the calculation, storing the result can yield substantial performance benefits. For example, if a derived metric like profit_margin is queried repeatedly while the underlying revenue and cost columns are not, pre-storing profit_margin can reduce the need for repeated computation and scanning.

```
-- Step 1: Create a new table with the pre-calculated column
CREATE OR REPLACE TABLE sales_data_optimized AS
SELECT
    order_id,
    revenue,
    cost,
    -- Pre-calculated intra-row value
    (revenue - cost) / NULLIF(revenue, 0) AS profit_margin
FROM sales_data;

-- Step 2: Query using the pre-calculated column
SELECT
    order_id,
    profit_margin
FROM sales_data_optimized
WHERE profit_margin > 0.2;
```

Intra-row calculations are an invaluable strategy for optimizing Snowflake workloads, but their implementation requires thoughtful consideration to balance performance gains with resource utilization.

Summary

This chapter outlined essential strategies for optimizing workloads in Snowflake to enhance performance and control costs. It explored techniques such as mitigating query and join skew, and architectural guidance for selecting the right data load processes based on data volume and transformation complexity. The chapter also introduced Snowpark for in-platform processing to reduce data movement and simplify infrastructure, and it emphasized the importance of designing the consumption layer for performance and cost-efficiency.

A real-world case study demonstrated how proactive monitoring, clustering, and tagging can prevent unexpected cost surges and ensure sustainable Snowflake operations.

As explained throughout this chapter, optimizing workloads in Snowflake requires a strategic balance of performance, cost, and operational efficiency. The next chapter looks at storage optimization, focusing on how data organization, compression, and lifecycle management can further reduce costs and improve performance across your Snowflake environment.

CHAPTER 7

Storage Optimization Strategies

Overview

In the previous chapters, we explored different ways to optimize and save costs on compute resources. This chapter focuses on storage optimization techniques. Snowflake has a compute and storage decoupled architecture, which makes it flexible to scale and charge both resources independently. While compute costs are often highlighted, storage can become a silent cost burster if not properly managed. In this chapter, we explore strategies to trace unused storage and design considerations that can be crucial in curtailing the underlying costs right at the starting line.

The strategies covered in this chapter include:

- Essential guidelines for maintaining a balance for a continuous data protection timeline.

- How much of an impact can reclustering make if not maintained well, and how can you tune it well to prevent exponentially growing costs?

- Using the temporary and transient tables to your advantage without exploding the underlying usage costs.

- Tuning the permanent tables and following a discipline in the cleanup process to control unnoticed storage costs.

And many more valuable scenarios and guidelines that can aid in keeping storage costs under control.

CHAPTER 7 STORAGE OPTIMIZATION STRATEGIES

Managing Database Storage Costs

Storage costs in Snowflake are calculated for data in any state, whether it's Active, Time Travel, or Fail-safe. Snowflake also adds internal storage usage to storage costs. Depending on the region Snowflake is hosted in, respective cloud providers charge external cloud storage used for data load and unload.

To manage database storage costs effectively, it's important to first understand the relevant parts of the storage architecture. Snowflake uses unlimited cloud storage for database and file stage storage, which cloud vendors promise as having "eleven nines" (99.999999999) of durability. All storage is stored redundantly in multiple availability zones (AZs).

Snowflake's "secret sauce" is its use of micro-partitions, which have the following characteristics:

- They are immutable file objects stored in blob storage.

- They are always compressed, encrypted, and typically 16MB in size, representing about 100-150MB of uncompressed data.

- When an UPDATE, DELETE, or MERGE operation occurs, a new micro-partition is created with a "date-modified" timestamp. Similarly, INSERT and COPY operations always create new micro-partitions.

- "Active" micro-partitions have no "date modified" timestamp, while "historical" micro-partitions have a modified timestamp for time-travel.

When a date modified timestamp is present, the continuous data protection (CDP) setting determines how long the data is available for recovery.

Understanding Continuous Data Protection (CDP)

Controlling storage costs in Snowflake is primarily driven by the continuous data protection (CDP) factor, which covers the total amount of data stored and the length of time the data is stored. CDP denotes the way of protecting data under any circumstances for a given period. Although the CDP is intended for recovery purposes (when required), it adds additional costs.

CDP, which includes Time Travel and Fail-safe, is a standard set of features available to all Snowflake accounts at no additional cost. However, since your account is charged for all data stored in tables, schemas, and databases, CDP can affect storage costs, depending on the total amount of data and how long it's stored, as illustrated in Figure 7-1.

Since data lifecycle states follow a sequence, any updated or deleted data protected by CDP will continue to incur storage costs until it is no longer in the Fail-safe state.

Time Travel enables:

- Accessing data that has been changed or deleted within a predefined period.

- Restoring accidentally or deliberately deleted databases, schemas, and tables using simple SQL.

- Cloning tables, schemas, and databases from key historical points.

- Backing up data for historical and archiving purposes through cloning.

- Querying data as it existed at previous points in time.

Fail-safe enables:

- Another seven days of recovery period after the Time Travel retention period.

Recovering data from Fail-safe requires Snowflake support intervention.

Snowflake charges the data that has been retained. It means that the charges are in place for active and as well inactive data associated with Time Travel and Fail-safe.

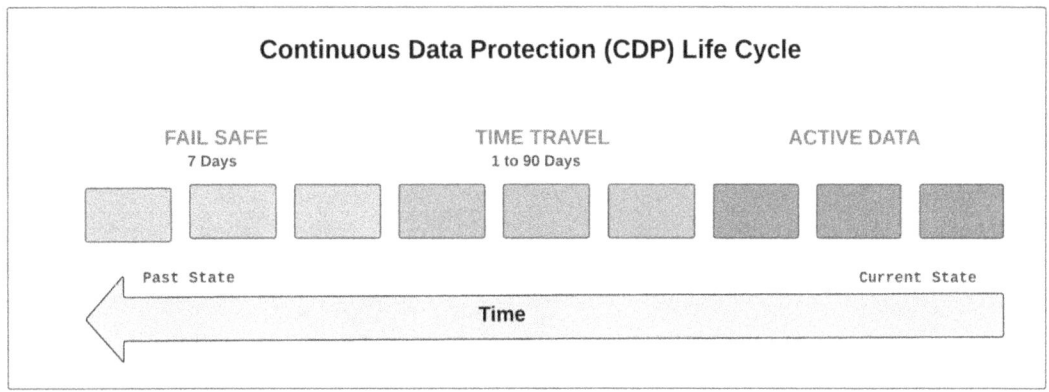

Figure 7-1. *The CDP lifecycle*

In real-time environments, the Time Travel and Fail-safe features are essential for ensuring data protection and recovery. However, these features come with significant storage costs. Balancing cost efficiency and data availability is crucial, and this is where storage optimization becomes vital. Before diving into effective optimization strategies, the following section explores the storage components that drive these costs.

Components of Snowflake Storage Costs

Snowflake Storage includes the following database objects and filesystems that are reflected in storage costs.

- Database tables, materialized views, and so on.
- Snowflake internal stage used for data loading/unloading. Files can be stored in compressed or uncompressed formats.
- Time Travel and Fail-safe for database tables.
- Clones of database tables that reference deleted data.

Usage is measured daily, and the monthly charge is calculated based on the average usage across the number of calendar days in the month. The monthly storage costs are determined by a flat rate per terabyte. The specific amount charged depends on the type of account and region. Time Travel and Fail-safe fees are calculated based on each 24-hour period, starting from when the data was last modified.

As shown in Table 7-1, table storage billing is calculated based on the volume of data stored in each CDP layer and the number of days the data is stored in a month.

Table 7-1. *Table Storage Billing Calculation*

Table Storage	Day 1 Table is created	Day 2 Every Record is Updated	Day 3-9 No Change	Day 10 Fail Safe Expired	No of Days	Avg. Storage Per Month
Active Table	1 TB	1 TB	1 TB	1 TB	10	0.3333
Time Travel (set to 1 day)	0 TB	1 TB	0 TB	0 TB	1	0.0333
Fail Safe (default 7 days)	0 TB	0 TB	1 TB	0 TB	1	0.0333
				Avg. Storage Per Month * Storage Cost ($23 /TB)		9.2000

Keep in mind that the size of the table shown is the number of bytes that will be scanned if the whole table is queried, but that can be different from physical bytes, especially for cloned tables and tables where data has been deleted. The table size shown does not include the space occupied by deleted data from a table, and the data is still stored in Snowflake until expiration of data in the Time Travel retention period (one day by default) and the Fail-safe period (seven days). The table size shown for these two periods is less than the actual physical bytes space occupied by the table.

Storage Optimization Strategies

Storage optimization in Snowflake is the process of storing data in a way that allows it to occupy less space, cost less, and work more efficiently. It involves techniques such as change data capture (CDC), data compression, cloning, and so on, to minimize storage space and lower costs. Implement the following guidelines for automatic data retention, archiving, and moving cold data to cloud storage to further optimize storage.

This section provides guidelines to improve data storage efficiency and increase overall system performance. The first section discusses the Time Travel cost with an example.

Time Travel and Fail-Safe

The appropriate use of Snowflake's data retention capabilities and the cleanup of unused tables are key to controlling storage costs. Managing the table lifecycle is crucial for controlling the cost implications of Time Travel and Fail-safe capabilities.

Storage calculation and charges:

- Storage is calculated and charged for data in Active, Time Travel, or Fail-safe states.

- Updated or deleted data protected by these features continue to incur storage costs until it leaves the Fail-safe state.

Extended Time Travel

The data retention setting determines the length of time historical data remains available, which in turn affects the charges. Data retention can be set at the account, database, schema, or table level via the DATA_RETENTION_TIME_IN_DAYS parameter.

Set data retention on database, schema, table using this statement:

```
--Set Data Retention time for Database
CREATE DATABASE DEMO_DB DATA_RETENTION_TIME_IN_DAYS = 10;
ALTER DATABASE DEMO_DB SET DATA_RETENTION_TIME_IN_DAYS=10;

--Set Data Retention time for Schema
CREATE SCHEMA DEMO_SCHM DATA_RETENTION_TIME_IN_DAYS=20;
ALTER SCHEMA DEMO_SCHM SET DATA_RETENTION_TIME_IN_DAYS=20;

--Set Data Retention time for Table
CREATE TABLE TPCH_SF1.CUSTOMER (C_CUSTKEY NUMBER, C_NAME VARCHAR) DATA_RETENTION_TIME_IN_DAYS=30;
ALTER TABLE TPCH_SF1.CUSTOMER SET DATA_RETENTION_TIME_IN_DAYS=30;
```

A value at a higher level (e.g., database) may be overridden by setting the value for an object at a lower level. In this case, 30 days of data retention for the table override the schema and database setting. The default is always one day, and transient and temporary tables may only be set to one or zero days. Historical micro-partitions are moved into Fail-safe when their data retention setting is exceeded and stored for seven days.

Managing the Cost of Time Travel

Consider the following questions to determine the appropriate data retention period:

1. Do you need point-in-time queries for changed historical data? Note that historical data is retained for a maximum of 90 days (for Enterprise Edition and above).

2. How far back will you need to restore to recover a dropped table?

3. How frequently do UPDATEs, DELETEs, and MERGEs run that modify data?

4. Do you need to generate historical clones for testing, debugging, and so on?

5. No backup is required for media failures, as data is stored redundantly in three availability zones, preserving eleven nines of durability.

6. Micro-partitions referenced by the clone will be retained until the expiration of all retention times in the referencing clone object is exceeded.

Fixing the Mistake with 90 Data Retention

This extreme example demonstrates Time Travel's effect on storage costs and why data retention management is critical.

Example 1:

- Create a permanent table with a 90-day data retention time (Table A).
- Populate it on day 1 with 10 GB.
- Drop Table A on day 1.
- Total cost = (90+7) * 10GB * price/GB/day.

This enables you to UNDROP the table for 97 days. However, you will be charged for 97 days of storage of 10 GB, even though you might never UNDROP and/or use it!

To Fix the data retention, assume that this table and its population were a mistake. Fortunately, depending on when the mistake is discovered, you can reduce the cost to a minimum of seven days for Fail-safe.

- UNDROP the table.
- Modify the data retention period to 0 days.
- Drop the table.

The data was previously moved into Fail-safe, so you are still subject to a minimum charge of seven days plus any time the mistake went undiscovered.

Example 2:

- Table A contains 1 GB of data and has a retention time of 30 days.
- It update all records once daily. Each day generates an additional 1 GB.
- This results in 30 GB after 30 days, including active and Time Travel bytes.

The cost of retained data varies depending on the data retention time and the amount of data changed daily. Users should adjust the retention time depending on loading patterns and how important each object (database, schema, or table) is to extend its history.

Manage Short-lived Permanent Tables

During ETL or data-modeling processes, short-lived tables are often created. For these tables, incurring the storage costs associated with CDP is unnecessary. By default, tables in Snowflake are created as permanent unless specified otherwise. Using temporary or transient tables can help save on Fail-safe and Time Travel costs.

Note Transient and temporary tables don't come with a Fail-safe period and can have a Time Travel retention period of either 0 or 1 day.

Temporary Tables

Temporary tables only exist for the lifetime of their associated session, after which data is purged and unrecoverable. Snowflake temporary tables have no Fail-safe and a Time Travel retention period of just 0 or 1 day. The Time Travel period ends as soon as the table is dropped.

Therefore, the maximum CDP charges for a temporary table are limited to one day (or less if the table is dropped manually or automatically when the session ends). During this time, Time Travel can be used on the table.

Transient Tables

Transient tables are unique to Snowflake and exist until explicitly dropped, after which historical data beyond the Time Travel retention period cannot be recovered. They have characteristics of both permanent and temporary tables:

- **Accessibility:** Unlike temporary tables, transient tables are not tied to a session and are accessible to all users with the proper permissions.

- **Persistence:** Similar to permanent tables, they remain available even after the session in which they were created ends.

As a result, the maximum CDP charges for a transient table are limited to one day. During this time, Time Travel can be used on the table. Figure 7-2 outlines the characteristics of different Snowflake table types, providing various options for interim data processing.

Table Type	Availability	Time-Travel Retention Period (days)	Fail-Safe period in Days
Temporary	Remainder of session	0 or 1 (Default is 1)	0
Transient	Until explicitly dropped	0 or 1 (Default is 1)	0
Permanent (Std Edition)	Until explicitly dropped	0 or 1 (Default is 1)	7
Permanent (Ent/ higher Edition)	Until explicitly dropped	0 to 90 (default is configurable)	7

Figure 7-2. Snowflake table types

Managing Large, High-churn Tables

To control storage and performance, it's essential to choose appropriate data protection strategies for high-churn dimension tables and low-churn fact tables, which consume a significant ratio of Time Travel and Fail-safe resources. Here are some optimization strategies:

- **Reduce the Time Travel limit:** Lowering the retention period can significantly reduce storage costs.

- **Rewrite applications to produce less churn:** For example, use `INSERT OVERWRITE` to minimize data modifications.

- **Investigate using hybrid tables:** Consider hybrid tables for specific use cases, such as high-volume logging

Note Evaluate the costs associated with using hybrid tables, especially when considering storage requirements and potential performance impacts on your workloads.

High-churn dimension tables can be identified by a high ratio of FAILSAFE_BYTES to ACTIVE_BYTES in the TABLE_STORAGE_METRICS view. In these scenarios, the costs and benefits of Time Travel and Fail-safe must be carefully weighed.

For low-durability data with a short lifespan, such as ETL worktables that only need to be persisted for less than a day, we recommend using the transient or temporary table types to reduce CDP costs.

Use the following query to trace Time Travel tables with high storage usage (the following image shows an example).

```
-- Top Time-Travel Tables
SELECT
    table_catalog,table_schema,table_name
    active_bytes,
    time_travel_bytes,
    failsafe_bytes,
    CASE WHEN active_bytes = 0 THEN 0 ELSE time_travel_bytes/active_bytes END as timetravel_ratio,
    CASE WHEN active_bytes = 0 THEN 0 ELSE failsafe_bytes/active_bytes END as failsafe_ratio
FROM snowflake.account_usage.table_storage_metrics
WHERE NOT deleted
AND active_bytes > 0
ORDER BY time_travel_bytes DESC
LIMIT 100;
```

	TABLE_CATALOG	TABLE_SCHEMA	TABLE_NAME	ACTIVE_BYTES	TIME_TRAVEL_BYTES	FAILSAFE_BYTES	TIMETRAVEL_RATIO	FAILSAFE_RATIO
1	DEMO_DB	TPCH_SF10	LINEITEM_CLONE	234171458	59785366	9852	6068.35	6068.35
2	DEMO_DB	TPCH_SF10	LINEITEM	563877490	5042106	23481	214.73	214.73

Use the following query to trace Fail-safe tables with high storage usage that include deleted tables (the following image shows an example).

```sql
-- Top Failsafe Tables (including deleted tables)
SELECT
    table_catalog,table_schema,table_name
    active_bytes,
    time_travel_bytes,
    failsafe_bytes,
    CASE WHEN active_bytes = 0 THEN 0 ELSE failsafe_bytes/active_bytes END as failsafe_ratio
FROM snowflake.account_usage.table_storage_metrics
WHERE active_bytes > 0
ORDER BY failsafe_bytes DESC
LIMIT 100;
```

TABLE_CATALOG	TABLE_SCHE	↑ TABLE_NAME	ACTIVE_BYTES	TIME_TRAVEL_BYTES	FAILSAFE_BYTES	TIMETRAVEL_RATIO	FAILSAFE_RATIO
1 DEMO_DB	TPCH_SF10	LINEITEM	563677490	5042106	23481	214.73	214.73

Storage Impact of Reclustering

When using cluster keys, it's important to be cautious as they can lead to micro-partitions aging and incurring unexpected storage costs. The reclustering process ensures that rows in a table are physically grouped according to the clustering key, creating new micro-partitions for the table. Even the inclusion of a small number of rows can result in the re-creation of all associated micro-partitions.

This process can cause significant data turnover since the original micro-partitions are deleted but retained for Time Travel or Fail-safe purposes. The original micro-partitions are only deleted after the Time Travel retention period (which can range from a minimum of one day up to 90 days) and the Fail-safe period (an additional seven days). This often leads to increased storage costs.

Recommendations:

> **Low data retention setting:** Consider setting a low data retention period when using cluster keys to minimize storage costs.
>
> **Periodic reviews:** Regularly review tables that are enabled for reclustering and monitor Time Travel and Fail-safe usage to manage and optimize storage costs effectively.

CHAPTER 7   STORAGE OPTIMIZATION STRATEGIES

Use the following query to trace storage usage of tables enabled for auto-clustering (the following image shows an example).

```
--Auto-clustering storage usage
WITH table_storage_cost AS (
SELECT
        id AS table_id,
        active_bytes,
        time_travel_bytes,
        failsafe_bytes,
        retained_for_clone_bytes,
        (active_bytes + time_travel_bytes + failsafe_bytes + retained_for_
clone_bytes) / POWER(1024, 4) AS total_storage_tb,
        total_storage_tb * 23.00 AS monthly_cost
--Adjust cost per TB per your billing
    FROM
        snowflake.account_usage.table_storage_metrics
    WHERE
        NOT deleted
        AND table_catalog <> 'SNOWFLAKE'
)
SELECT t.table_catalog,t.table_name,t.table_schema,t.clustering_key,t.auto_
clustering_on,
        s.active_bytes,
        s.time_travel_bytes,
        s.failsafe_bytes,
        s.total_storage_tb,
        s.monthly_cost
FROM snowflake.account_usage.tables t JOIN table_storage_cost s
ON t.table_id = s.table_id
WHERE   t.deleted IS NULL
        AND t.table_catalog <> 'SNOWFLAKE'
        AND t.auto_clustering_on = 'YES';
```

TABLE_CATALOG	TABLE_NAME	TABLE_SCHEMA	CLUSTERING_KEY	AUTO_CLUSTERING_(	TOTAL_STORAGE_TB	MONTHLY_COST
DEMO_DB	PART	TPCH_SF10	LINEAR(P_TYPE)	YES	5	115
DEMO_DB	ORDERS	TPCH_SF10	LINEAR(O_CUSTKEY, TO_DATE(O_ORDERDATE))	YES	5	115
DEMO_DB	LINEITEM	TPCH_SF10	LINEAR(L_PARTKEY)	YES	7.01	161.23

Point-in-Time Cloning

Snowflake's zero-copy cloning provides a convenient way to "snapshot" any table, schema, or database and create a derived copy that initially shares the underlying storage. This method facilitates instant backups without incurring additional costs.

Each clone has its own separate lifecycle, and changes can be made to the original object or clone independently. There are no limits on the number of clones that can be created, resulting in an *n*-level hierarchy of cloned objects with shared and independent storage. CDP safeguards new micro-partitions that are exclusively owned by the clone, as each change to the clone generates new micro-partitions. Figure 7-3 illustrates the underlying architecture of a cloned table, which doesn't require additional space when a point-in-time table is cloned.

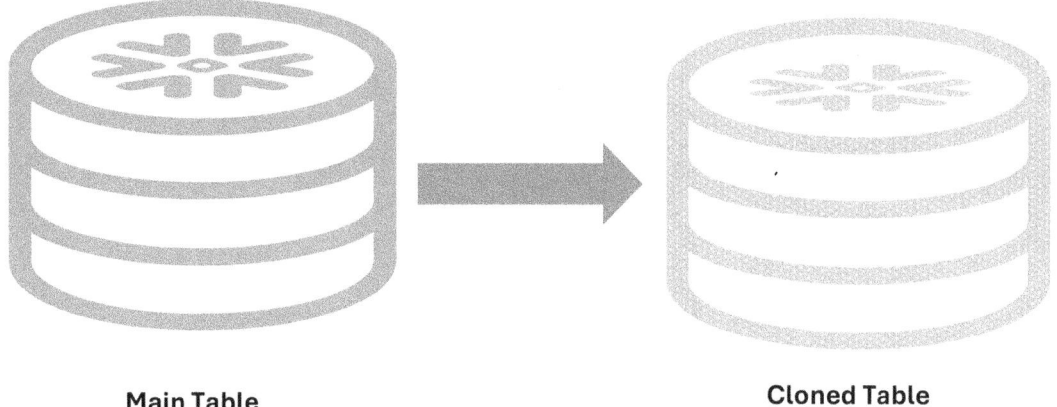

Main Table   Cloned Table

Figure 7-3. *Snowflake table cloning*

Cloned Data Storage

A cloned table does not use additional storage until rows are added to or modified/deleted from the table. The displayed table size may be larger than the actual physical bytes stored for the table; as a result, the cloned table contributes less to the overall storage than the size indicates.

How to Determine Storage Based on Cloned Data

Cloning data incurs no additional costs until the data is modified or deleted. This process is quick and efficient, benefiting all users through *zero-copy cloning*.

To determine the exact percentage of your table storage that comes from cloned data, leverage the CLONE_GROUP_ID column in TABLE_STORAGE_METRICS using following query (the following image shows an example).

```
WITH tab_storage AS (
 SELECT
    TABLE_NAME,
    clone_group_id,
    sum(retained_for_clone_bytes) AS clone_bytes,
    sum(active_bytes) + sum(time_travel_bytes) AS referred_bytes
 FROM snowflake.account_usage.table_storage_metrics
 WHERE active_bytes > 0
 AND TABLE_NAME  IN ( 'LINEITEM')
 AND TABLE_SCHEMA = 'TPCH_SF10'
 GROUP BY ALL
 )
SELECT cs.*, referred_bytes / clone_bytes AS ratio
FROM tab_storage cs
WHERE clone_bytes > 0 AND ratio > 1
ORDER BY clone_bytes DESC;
```

	TABLE_NAME	CLONE_GROUP_ID	CLONE_BYTES	REFERRED_BYTES	RATIO
1	LINEITEM	4102	15131426131	45449011712	3.003617

You can get an idea of how much of the original data is being "referenced" by the clone from the ratio shown in the previous output. So, if you create a clone table, in general, the new cloned table is assigned the original table's CLONE_GROUP_ID. Your CLONE_BYTES value is updated and billed to this table, which is retained after deletion because one or more clones reference it.

Manage Storage in Snowflake Stages

Data files staged in Snowflake internal stages are not subject to the additional costs associated with Time Travel and Fail-safe, but they do incur standard data storage costs. To help manage these costs, it is best to monitor these files and remove them from the stages once the data has been loaded and the files are no longer needed. This can be done as part of the COPY INTO command or via the REMOVE command after loading.

The account usage view provides two views related to stages:

- **STAGES:** Lists all the stages defined in your account but does not show how much storage each stage consumes.

- **STAGE_STORAGE_USAGE_HISTORY:** Shows the total usage of all stages for the last six months, including named internal stages, but does not provide detailed usage information.

To gain additional insight into the storage of each stage, users must use the LS command at each stage and process the result to calculate the total size. Files can then be removed using the REMOVE command.

Alternatively, Snowflake scripting can be used to get a collective insight into each stage's total size. Note that this script may take time based on the number of files in the stage location.

This query retrieves internal stage storage usage (the following image shows an example):

```
show stage;
list @demo_int_stage;
select SUBSTR($1,1,POSITION('\/',$1,1)-1) AS stage_name,
       ROUND(SUM($2)/POWER(1024, 3),2) AS total_storage_gb
from table(result_scan(last_query_id()))
group by all;
```

	STAGE_NAME	TOTAL_STORAGE_GB
1	demo_int_stage	28.29

Delete Files Older Than a Specific Date in a Snowflake Stage

From time to time, you may need to delete old files stored in a Snowflake stage. The REMOVE command can filter and delete files using a regular expression, but it does not support filters based on date. As a workaround, it's possible to write a stored procedure script to parse the output of the LIST command and generate the required REMOVE commands to delete files older than a specific date (such as 14 days).

Here's a stored procedure example (the following image shows an example):

```
CREATE OR REPLACE PROCEDURE remove_stage_files( stage_name VARCHAR, num_days number, dry_run boolean )
RETURNS varchar
LANGUAGE sql
EXECUTE AS CALLER
AS
DECLARE
    ListFiles RESULTSET;
    LastModified DATE;
    RemovedCount NUMBER := 0;
    TotalCount NUMBER := 0;
BEGIN
    ListFiles := (EXECUTE IMMEDIATE 'LS @' || stage_name );
    LET C1 CURSOR FOR ListFiles;
    FOR files IN C1 DO
        TotalCount := TotalCount + 1;
        LastModified := TO_DATE(LEFT( files."last_modified", LENGTH(files."last_modified") - 4 ), 'DY, DD MON YYYY HH24:MI:SS' );
        IF (LastModified <= DATEADD( 'day', -1 * num_days, current_timestamp())) THEN
            RemovedCount := RemovedCount + 1;
            IF (NOT dry_run) THEN
                EXECUTE IMMEDIATE 'RM @' || files."name";
            END IF;
        END IF;
    END FOR;
```

```
    RETURN RemovedCount || ' of ' || TotalCount || ' files ' || IFF(dry_
run,'will be','were') || ' deleted.';
END;
```

Stored procedure run output:

```
CALL remove_stage_files( 'demo_stage', 7, true );
```

> **REMOVE_STAGE_FILES**
>
> 1   309 of 353 files will be deleted.

This stored procedure:

- **Lists files:** Retrieves the list of files in the specified stage.

- **Checks the last modified date:** Parses the last modified date of each file.

- **Deletes old files:** Generates and executes `REMOVE` commands for files older than the specified number of days.

- **Includes a dry run option:** Allows for a dry run to see how many files would be deleted without actually removing them.

By using this procedure, you can efficiently manage and delete old files in Snowflake stages based on their last modified date.

Data/Table Cleanup Based on Usage

One of the best ways to control storage costs and optimize data in Snowflake is to periodically delete aging data that exceeds the minimum threshold. Businesses can minimize data storage by identifying and eliminating outdated or unnecessary data, thereby enhancing performance and reducing costs.

Snowflake provides the ability to automate this process using tasks or custom lifecycle policies tailored to the specific needs of the business. These policies can be enforced against permanent tables, internal stages, managed iceberg tables, and hybrid tables. Businesses can automate data retention and deletion using Snowflake tasks or custom scripts to ensure efficient data management that aligns with their specific use case and retention requirements.

Table Cleanup

Unused tables increase storage costs, so it is beneficial to perform periodic table analysis and delete unused tables or purge unwanted data. Before dropping tables, identify if the table is present for disaster recovery or shared with other Snowflake accounts, which could explain why it is not accessed often.

Storing data and ingesting new data into Snowflake without it being read wastes resources. Consider deleting these tables to save on storage costs or stop writing new data to save on credits consumed by ingestion.

- Tables that are never queried
- Tables in which data is written but not read

Here's an example query for table cleanup (the following image shows an example):

```
with table_storage_cost AS (
    SELECT
        id AS table_id,
        id != clone_group_id AS is_cloned,
        table_catalog,
        table_schema,
        table_name,
        active_bytes,
        time_travel_bytes,
        failsafe_bytes,
        retained_for_clone_bytes,
        (active_bytes + time_travel_bytes + failsafe_bytes + retained_for_clone_bytes) / POWER(1024, 4) AS total_storage_tb,
        total_storage_tb * 23.00 AS monthly_cost --Adjust cost per TB per your billing
    FROM
        snowflake.account_usage.table_storage_metrics
```

```sql
    WHERE
        NOT deleted
        AND table_catalog <> 'SNOWFLAKE'
),
table_dml_details AS (
    SELECT
        objects_modified.value:objectId::INTEGER AS table_id,
        COUNT(*) AS dml_count,
        MAX(query_start_time) AS dml_last_ts,
        TIMEDIFF(DAYS, dml_last_ts, CURRENT_TIMESTAMP()) AS dml_days_past
    FROM
        snowflake.account_usage.access_history,
        LATERAL FLATTEN(snowflake.account_usage.access_history.objects_modified) AS objects_modified
    WHERE
        objects_modified.value:objectDomain::TEXT = 'Table'
        AND table_id IS NOT NULL
        AND query_start_time > DATEADD('days', -30, CURRENT_TIMESTAMP)
    GROUP BY
        table_id
), table_ddl_details AS (
    SELECT
        object_modified_by_ddl:objectId::INTEGER AS table_id,
        COUNT(*) AS ddl_count,
        MAX(query_start_time) AS ddl_last_ts,
        TIMEDIFF(DAYS, ddl_last_ts, CURRENT_TIMESTAMP()) AS ddl_days_past
    FROM
        snowflake.account_usage.access_history
    WHERE
        object_modified_by_ddl:objectDomain::TEXT = 'Table'
        AND table_id IS NOT NULL
        AND query_start_time > DATEADD('days', -30, CURRENT_TIMESTAMP)
    GROUP BY
        table_id
),
```

```
table_access_details AS (
    SELECT
        objects_accessed.value:objectId::INTEGER AS table_id,
        -- objectId is null for secured views or tables from a data share
        COUNT(*) AS access_count,
        MAX(query_start_time) AS access_last_ts,
        TIMEDIFF(DAYS, access_last_ts, CURRENT_TIMESTAMP()) AS access_days_past
    FROM
        snowflake.account_usage.access_history,
        LATERAL FLATTEN(snowflake.account_usage.access_history.base_objects_accessed) AS objects_accessed
    WHERE
        objects_accessed.value:objectDomain::TEXT = 'Table'
        AND table_id IS NOT NULL
        AND query_start_time > DATEADD('days', -30, CURRENT_TIMESTAMP)
    GROUP BY
        table_id
)
SELECT
    table_catalog,
    table_schema,
    table_name,
    active_bytes,
    time_travel_bytes,
    failsafe_bytes,
    retained_for_clone_bytes,
    total_storage_tb,
    monthly_cost,
    dml_count,
    dml_last_ts,
    dml_days_past,
    ddl_count,
    ddl_last_ts,
    ddl_days_past,
```

```
        access_count,
        access_last_ts,
        access_days_past
FROM
    table_storage_cost strg
LEFT JOIN
    table_dml_details dml ON strg.table_id = dml.table_id
LEFT JOIN
    table_ddl_details ddl ON strg.table_id = ddl.table_id
LEFT JOIN
    table_access_details sel ON strg.table_id = sel.table_id
--Uncomment to find the tables that were not accessed in last 90 days
--WHERE sel.access_days_past is null
--   AND ddl.ddl_days_past is null
--   AND dml.dml_days_past is null
ORDER BY access_days_past, monthly_cost DESC
;
```

TABLE_SCHEMA	TABLE_NAME	TOTAL_STORAGE_TB	MONTHLY_COST	DML_COUNT	DML_LAST_TS	DML_DAYS_PAST	DDL_COUNT	DDL_LAST_TS	DDL_DAYS_PAST	ACCESS_COUNT	ACCESS_LAST_TS	ACCESS_DAYS_PAST
TPCH_SF100	LINEITEM	0.1047	2.4084	1	2024-11-16 08:55:21.753	9	1	2024-11-16 08:53:24.594 -0800	9	null	null	null
TPCH_SF100	ORDERS	0.0946	2.1758	1	2024-11-16 08:59:03.736	9	1	2024-11-16 08:55:09.097 -0800	9	null	null	null
TPCH_SF100	PARTSUPP	0.0934	2.1481	1	2024-11-16 09:00:23.545	9	1	2024-11-16 09:00:10.419 -0800	9	2	2024-11-18 05:29:46.661 -0800	7

This query helps identify tables that are not frequently accessed or modified, allowing you to make informed decisions about which tables to delete or stop writing new data to.

Manage the Storage Lifecycle

Most organizations store cold data much longer than required to comply with regulations, leading to expensive storage costs. To mitigate Snowflake storage costs, consider using cheaper cloud storage for cold data.

CHAPTER 7 STORAGE OPTIMIZATION STRATEGIES

While hot data is retained in Snowflake-managed tables, cold data can periodically be offloaded to an attached cloud blob store (where intelligent-tiering is likely enabled by enterprise policy). After moving cold data to cloud blob storage, create an external table and a view to allow users to access both hot and cold data until it is purged. Intelligent-tiering helps save costs and meet governance requirements by automatically expiring and deleting data beyond its retention period.

Here's an example process:

1. Unload data to an external stage:

```
--UNLOAD data to External stage DEMO_EXT_STAGE
COPY INTO @DEMO_EXT_STAGE
FROM (SELECT l_orderkey, l_partkey, l_suppkey, l_quantity, l_shipdate, l_receiptdate
        FROM LINEITEM
    WHERE l_shipdate < '2019-01-01')
PARTITION BY ('date=' || to_varchar(l_shipdate, 'YYYY-MM-DD'))
FILE_FORMAT = (TYPE=parquet)
MAX_FILE_SIZE = 32000000
HEADER=true;
```

2. Delete the old data:

```
--Delete old data
DELETE FROM TPCH_SF100.LINEITEM WHERE l_shipdate < '2019-01-01';
```

3. Create an external table pointing to the cloud blob storage:

```
--Create External table poinintg to cloud blob stoage
CREATE OR REPLACE EXTERNAL TABLE TPCH_SF100.EXT_LINEITEM (
    l_orderkey int AS (value:l_orderkey::int),
    l_partkey int AS (value:l_partkey::int),
    l_suppkey int AS (value:l_suppkey::int),
    l_quantity int AS (value:l_quantity::int),
    l_shipdate date AS TO_DATE(value:l_shipdate::date),
    l_receiptdate date AS TO_DATE(value:l_receiptdate::date)
)
```

CHAPTER 7 STORAGE OPTIMIZATION STRATEGIES

```
PARTITION BY (l_shipdate) location=@DEMO_EXT_STAGE/
FILE_FORMAT = (type=parquet);
```

4. Create a unified view to access hot and cold data:

```
--Creatge unified view to access hot and cold data
CREATE OR REPLACE VIEW ALL_LINEITEM AS
SELECT l_orderkey, l_partkey, l_suppkey, l_quantity, l_shipdate,
l_receiptdate FROM TPCH_SF100.LINEITEM
UNION
SELECT l_orderkey, l_partkey, l_suppkey, l_quantity, l_shipdate,
l_receiptdate FROM TPCH_SF100.EXT_LINEITEM;
```

5. Query the recent data:

```
--When recent data is accessed, efficient pruning is seen in the
query profile
SELECT COUNT(*) from ALL_LINEITEM WHERE l_shipdate >= TO_
DATE('2019-01-01');
```

Query Profile Output:
Show the query only accessed fewer partitions eliminating partitions from the external table

Statistics

Scan progress	100.00%
Bytes scanned	838.69MB
Percentage scanned from ...	0.00%
Bytes sent over the network	1.10GB
Partitions scanned	154
Partitions total	534

By following this process, you can efficiently manage the storage lifecycle, reduce costs, and ensure compliance with data retention policies.

Workspace Governance and Cleanup

It's crucial to establish best practices for managing workspace (data lab) tables to avoid storing unnecessary data/tables longer than required while ensuring compliance and supporting business continuity. As illustrated in Figure 7-4, key practices include defining retention policies, automating cleanup processes, and implementing time-based deletion strategies to effectively manage the entire lifecycle of tables.

Consider these key practices:

- **Set clear retention policies:** Workspace tables are typically temporary. Establish how long the data should remain in the workspace before it becomes stale or unnecessary.

- **Automate table deletion:** Set automated jobs or scripts to clean up workspace tables after a certain period (e.g., 30, 60, or 90 days) or once the data is no longer required for analysis.

- **Focus on user awareness:** Ensure that users are aware of workspace table policies and regularly clean up their own data.

- **Implement data lifecycle management:** If historical data might be needed later, consider archiving the workspace data before deleting it. You could offload the data to another database or cloud storage system

- **Move to production:** For tables transitioning from workspace to production, ensure they are cleaned up and moved correctly (e.g., renaming, reformatting, or transforming them into a production schema).

- **Log cleanup activities:** Keep logs of all workspace table cleanup operations for auditing and troubleshooting. Include information such as the tables deleted, reasons for deletion, and the automated process rule that initiated the cleanup.

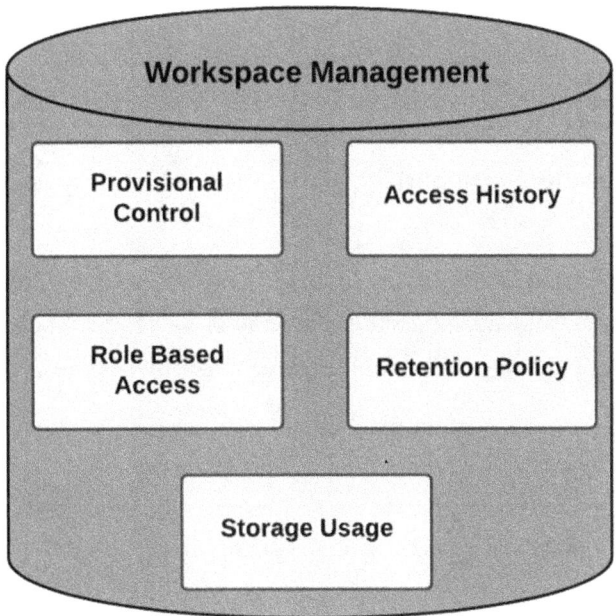

Figure 7-4. Workspace management

The Data Lakehouse, a Distributed Architecture

While many customers value the simplicity of Snowflake's fully managed storage, some prefer to manage their own storage using open table formats supported by Apache Iceberg tables. The data lakehouse architecture combines the scalability and flexibility of data lakes with the governance, schema enforcement, and transactional properties of data warehouses using Iceberg tables. As shown in Figure 7-5, detaching storage from compute enables the option to plug in a compute engine based on application requirements.

Iceberg Tables: An Emerging Architecture in Lakehouse Design

Iceberg tables bring the easy management and great performance of Snowflake to data stored externally in an open format, making them a key component of the emerging Lakehouse architecture. They also make onboarding easier and cheaper without requiring upfront ingestion. To give customers flexibility in how they fit Snowflake into their architecture, Iceberg tables can be configured to use either Snowflake or an external service like AWS Glue as the table's catalog to track metadata. A simple one-line SQL command can convert the table to Snowflake in a metadata-only operation.

CHAPTER 7　STORAGE OPTIMIZATION STRATEGIES

Advantages:

- **Improved query performance:** On average, at least two times better than external tables.

- **Write operations:** Support for INSERT, MERGE, UPDATE, and DELETE operations.

- **Feature compatibility:** Many features work seamlessly, including data sharing, role-based access controls, Time Travel, Snowpark, object tagging, row access policies, and masking policies.

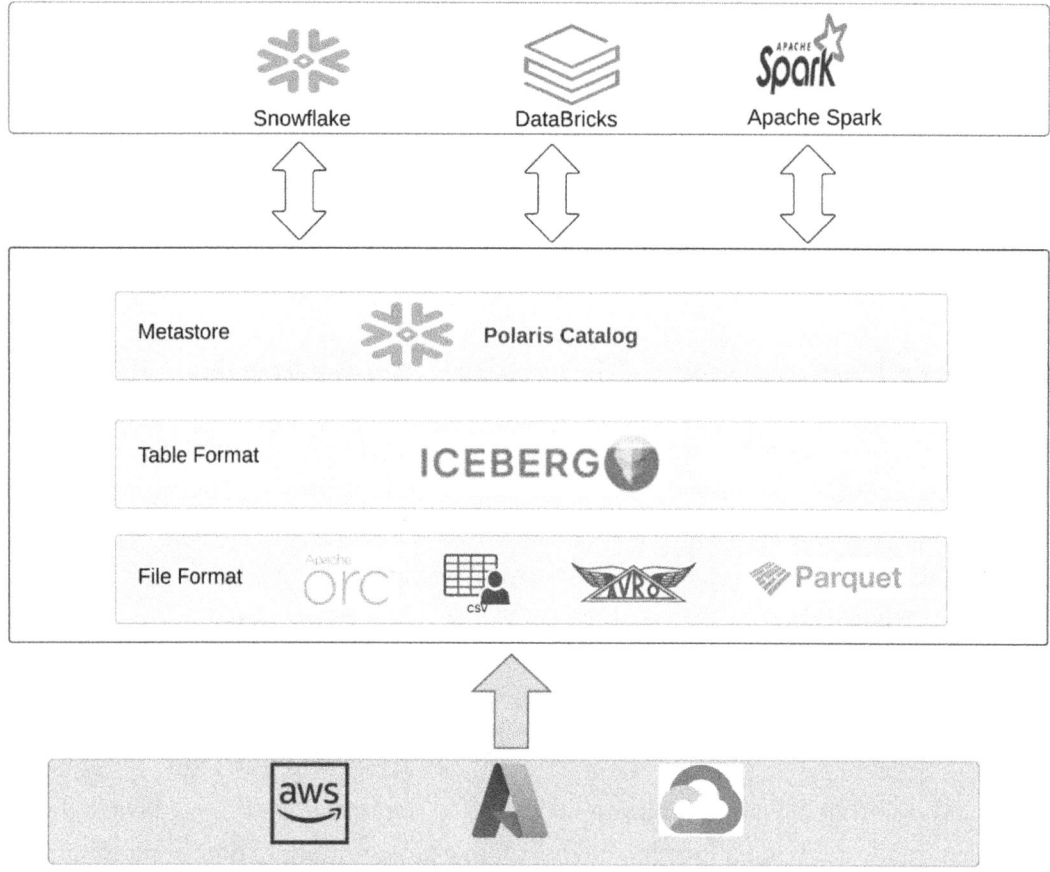

Figure 7-5. Data lakehouse

By leveraging Iceberg tables, customers gain flexibility in integrating Snowflake into their architecture, enhancing both performance and manageability of their data.

CHAPTER 7 STORAGE OPTIMIZATION STRATEGIES

Profiling Storage and Cost Use

Having explored various storage optimization strategies, in this section, we explain the factors contributing to storage costs and how to track usage. Users with the ACCOUNTADMIN role can view the amount of data stored either via the UI (the Storage filter under the Usage Type drop-down) or by executing queries against the ACCOUNT_USAGE and ORGANIZATION_USAGE views, which are listed in Table 7-2.

Table 7-2. Storage Usage Views

View Name	Details
HYBRID_TABLES	Shows data storage in bytes for each hybrid table row in the account.
USAGE_IN_CURRENCY_DAILY	Shows details on the daily average storage in bytes along with the cost of that usage in the organization's currency.
TABLE_STORAGE_METRICS	Shows details on storage in bytes for tables, including storage that is no longer active but continues to incur costs (e.g., deleted tables with the Time Travel retention period).
STAGE_STORAGE_USAGE_HISTORY	Shows the average daily storage usage, in bytes, for all the Snowflake stages, including named internal stages and default staging areas.
DATABASE_STORAGE_USAGE_HISTORY	Shows the average daily storage in bytes for each database, including data in Time Travel in the account/organization.
STORAGE_DAILY_HISTORY	Shows the average daily storage for storage in bytes. It combines database storage (DATABASE_STORAGE_USAGE_HISTORY) and stage storage (STAGE_STORAGE_USAGE_HISTORY).
LISTING_AUTO_FULFILLMENT_DATABASE_STORAGE_DAILY	Shows data storage in bytes for databases fulfilled to other regions by cross-cloud auto-fulfillment.

Storage pricing is based on the average terabytes per month of all customer data stored in the Snowflake account. The average terabytes per month is calculated by taking an hourly snapshot of all customer data and then averaging this across each day. This daily average is displayed in the Snowflake service, and the monthly charge is based on the average calculated across the number of calendar days in that month. If customer data is compressed, the compressed file size is used to calculate the total storage used.

Table Usage Metrics

The TABLE_STORAGE_METRICS view provides detailed information, including a breakdown of the physical storage (in bytes) for table data in the states of the CDP lifecycle listed in Table 7-3.

Table 7-3. Table Usage Metrics

Column Name	Details
ACTIVE_BYTES	Number of bytes in active micro-partitions
TIME_TRAVEL_BYLES	Number of bytes in historical micro-partitions
FAILSAFE_BYTES	Number of bytes in Fail-safe
RETAINED_FOR_CLONE_BYTES	Number of bytes in micro-partitions referenced by clones
TABLE_DROPPED	Date and time when the table was dropped
TABLE_ENTERED_FAILSAFE	Date and time the table entered Fail-safe; requires Snowflake support to restore

Storage Profiling Using UI

The Cost Management interface provides a comprehensive toolkit and insights to manage your cash flow. It offers complete visibility at the organization and account levels, providing the insights needed to make timely decisions and optimize costs proactively. Figure 7-6 illustrates the top databases by storage usage.

Figure 7-6. Database storage usage

Storage Profiling Using Pre-Built Queries

To gain detailed historical information about storage usage in your Snowflake account, use the suggested queries for more granular insights. While comparing billed storage credits against usage views, slight differences may be observed, mainly for the DATABASE_STORAGE_USAGE_HISTORY view.

What is the best view to show the matching result? The STORAGE_USAGE view takes the total size of all data files containing the account data. This tracks the amount of physical storage spent and refers to different statistics. The same information is also fetched in the UI.

Here's an example query for database storage usage (the following image shows an example):

```
-- Database Storage Usage
SELECT usage_date,
       month(usage_date) as usage_month,
       (sum(storage_bytes))::number(16,2) as storage_bytes,
       (sum(stage_bytes))::number(16,2) as stage_bytes,
       (sum(failsafe_bytes))::number(16,2) as failsafe_bytes,
       (sum(storage_bytes+stage_bytes+failsafe_bytes)/POWER(1024, 3)
)::number(12,2) total_GB
FROM SNOWFLAKE.ACCOUNT_USAGE.STORAGE_USAGE
WHERE usage_date >= '2024-11-25'
GROUP BY month(usage_date),usage_date
ORDER BY usage_date;
```

CHAPTER 7   STORAGE OPTIMIZATION STRATEGIES

	USAGE_DATE	USAGE_MONTH	STORAGE_BYTES	STAGE_BYTES	FAILSAFE_BYTES	TOTAL_GB
1	2024-11-25	11	25361408512.00	2636961172.00	119287.00	26.08
2	2024-11-26	11	36092912209.00	4104246499.00	115712.00	37.44
3	2024-11-27	11	46086217607.00	22264792570.00	8014656600.00	71.12
4	2024-11-28	11	33938432000.00	37895738111.00	20222459392.00	85.73
5	2024-11-29	11	34761349890.00	37895716227.00	20222439715.00	86.50
6	2024-11-30	11	56565228921.00	29613025963.00	15395505703.00	94.60

To find average daily storage usage (in bytes) and credits consumed for all accounts, query the STORAGE_DAILY_HISTORY view located in the ORGANIZATION_USAGE schema. The output projects credits used by database storage and stages (e.g., data in Time Travel, Fail-safe, internal stages, etc.).

Here's an example query for average daily storage usage (the following image shows an example):

```
SELECT
    SERVICE_TYPE,
    ACCOUNT_NAME,
    MONTH(USAGE_DATE) ||'-'|| YEAR(USAGE_DATE) AS MM_YYYY,
    REGION,
    ((avg(AVERAGE_BYTES)/POWER(1024, 4)))::NUMBER(12,2) STORAGE_TB,
    SUM(CREDITS)::NUMBER(12,2) AS CREDITS
FROM "SNOWFLAKE"."ORGANIZATION_USAGE"."STORAGE_DAILY_HISTORY"
WHERE SERVICE_TYPE= 'STORAGE'
GROUP BY ALL;
```

	SERVICE_TYPE	ACCOUNT_NAME	MM_YYYY	REGION	STORAGE_TB	CREDITS
1	STORAGE	PG56314	11-2024	AWS_US_EAST_2	36.91	400.33

By using these queries, you can gain detailed insights into storage usage and optimize costs effectively.

Summary

By utilizing the Snowflake Cost Management UI, Account Usage, and the Information Schema functions and views described in this chapter, you can effectively monitor your data storage. This, in turn, helps you keep storage costs under control and understand how your storage is growing over time. Investing in a new approach to storage management offers an opportunity to significantly improve your return on investment (ROI) with Snowflake.

Automating some of the best practices discussed in this chapter can help you identify outliers and storage usage trends. This proactive approach will enable you to formulate usage insights and take the necessary actions to optimize storage costs.

In the next chapter, we explore effective data access patterns, leveraging Snowflake's acceleration features to enhance data modeling optimization.

CHAPTER 8

Data Modeling Optimization Strategies

Overview

Optimizing data models is essential for achieving high performance and cost efficiency in Snowflake. This chapter examines practical strategies for aligning data modeling with query workloads, enhancing storage and retrieval efficiency, and leveraging Snowflake's advanced features. By understanding your data access patterns, you can select the appropriate modeling style—whether star, snowflake, or hybrid—and apply targeted optimizations, such as clustering keys, materialized views, and search optimization.

This chapter explores practical strategies for leveraging Snowflake's acceleration features to build high-performance, optimized data models. The primary goal is to enhance warehouse efficiency by reducing data movement (I/O) and improving query execution times. A well-designed physical data model effectively organizes data for optimal storage and retrieval.

Key modeling optimization strategies covered in this chapter include:

- Structuring data in a logical, query-friendly format to improve scan efficiency and reduce compute overhead.

- Using materialized views to cache complex query results and avoid repeated full-table scans, especially for frequently accessed aggregations.

- Enabling search optimization on columns used in selective filters to accelerate point-lookups and improve performance on large datasets.

- Using hybrid tables to support both real-time transactional updates and analytical queries in a single Snowflake table.

- Applying best practices for working with formats like JSON, including flattening nested structures and indexing key attributes.

- Taking advantage of Snowflake's flexible architecture to support complex use cases like dynamic schemas, evolving data structures, and multi-tenant environments.

With these foundational strategies in place, we now turn to clustering keys, a powerful technique for improving query performance by organizing data to reduce scan costs and enhance pruning efficiency.

Clustering Keys: A Foundation for Performance

We begin with one of Snowflake's most impactful optimization techniques—clustering keys. Clustering plays a central role in reducing query latency and compute resource usage by organizing micro-partitions to improve pruning efficiency. Effective data organization is critical for efficient query pruning, making clustering a fundamental aspect of performance optimization in Snowflake. When used strategically, clustering enables Snowflake to read only the relevant slices of data, resulting in faster execution and lower costs.

Here is how clustering can optimize data modeling:

- **Maximize pruning for better query performance:** Efficiently filter out irrelevant data to improve query execution speed.

- **Utilize natural clustering with ORDER BY when possible:** Leverage natural data order to optimize data storage and retrieval.

- **Optimize large tables with frequent filtering or range queries:** Implement strategies to efficiently handle common query patterns on large datasets.

- **Balance the benefits and costs of clustering**: Carefully consider the tradeoffs between clustering benefits and potential drawbacks.

- **Clustering considerations:**
  - Avoid clustering on rapidly changing tables (consider DML workload)
  - Avoid using columns with extremely high or low cardinality (e.g., timestamp, gender)
  - Clustering on VARCHAR columns (first six bytes are used)

In Snowflake, data is naturally clustered based on the order in which data load operations are executed. This order enables efficient pruning performance when filtering by attributes such as dates or unique identifiers, especially if they are time-related. However, if queries filter by attributes other than those used for clustering, pruning performance can degrade. In such cases, the system may need to read every micro-partition of the table, leading to slower query execution.

Now that we have covered the fundamentals of clustering, we next examine the cost implications of enabling automatic clustering in Snowflake.

Estimate Cost of Enabling Automatic Clustering

Before enabling automatic clustering, it's essential to understand the operational implications. While this feature can significantly enhance query performance, it also introduces ongoing compute and potential storage costs. This section provides a detailed breakdown of these cost components.

Compute Costs

Snowflake leverages serverless compute resources to cluster a table initially. It also utilizes these resources to maintain the table in an optimally clustered state as new data is added. The frequency of changes to the table directly impacts maintenance costs; more frequent modifications result in high maintenance overhead.

Storage Costs

Automatic clustering typically does not incur additional storage costs, as it reorganizes the existing data. However, if reclustering results in an increase in the size of Time Travel and Fail-safe storage, it could potentially lead to higher storage costs.

Estimating the Impact

The costs associated with automatic clustering depend on several factors, including the cardinality of the clustering key, the table size, and the DML patterns. Predicting the cost impact of changes to the clustering key can be a complex process. Understanding these cost implications is crucial for designing and optimizing efficient databases.

To estimate the cost of automatic clustering, you can utilize the ESTIMATE_AUTOMATIC_CLUSTERING function.

1) Estimate the cost of clustering a table for the first time.

2) Estimate the cost of changing the cluster key of a table.

3) When feasible, estimate the maintenance costs associated with a table after it has been clustered around the designated key.

> **Note** In some cases, a more detailed DML history may be necessary to accurately predict future maintenance expenses.

Cost estimates for automatic clustering are generated by sampling a subset of micro-partitions within the table and recording the clustering execution time. The most frequent cause of inaccurate maintenance cost estimations arises from a mismatch between the past DML patterns used for the estimate and the actual future DML patterns. If the function cannot determine a maintenance cost estimate, Snowflake provides a reason for this in the output.

Table 8-1 lists examples of estimate outputs and potential reasons for failure.

Table 8-1. Automatic Clustering Cost Estimate

Estimate Component	Definition	Factors That Influence Cost
Initial cost	The one-time cost to recluster data, based on the new cluster key	Cluster key cardinality, the degree to which the data was already well-clustered relative to the new clustering key and table size
Maintenance	This is the ongoing cost to maintain the data in a well-clustered state	Clustering key cardinality, DML patterns (write, update, etc.)

Consider this clustering costs estimate example:

```
SELECT SYSTEM$ESTIMATE_AUTOMATIC_CLUSTERING_COSTS('LINEITEM',
'(L_SHIPDATE)');
{
  "reportTime": "Tue, 17 Dec 2024 03:47:02 GMT",
  "clusteringKey": "LINEAR(L_SHIPDATE)",
  "initial": {
    "unit": "Credits",
    "value": 98.2,
    "comment": "Total upper bound of one-time cost"
  },
  "maintenance": {
    "unit": "Credits",
    "value": 10.0,
    "comment": "Daily maintenance cost estimate provided based on DML
    history from the
    past seven days."
  }
}
```

These estimates provide the clarity needed to move forward with enabling the clustering key on the shipment date and utilizing it in the predicate clause to enhance performance.

```
ALTER TABLE LINEITEM CLUSTER BY (L_SHIPDATE);
SELECT * FROM LINEITEM WHERE L_SHIPDATE='2024-12-01';
```

For tables with clustering enabled, it's important to monitor DML activity closely. Frequent updates, inserts, or deletes, especially on clustering key columns, can significantly increase maintenance costs due to the overhead of reclustering.

Track Updates on Clustered Tables

To assess the impact of these operations, use the TABLE_DML_HISTORY view to track the volume and type of DML executed. Additionally, the AUTOMATIC_CLUSTERING_HISTORY view provides insights into how these changes affect credit consumption for automatic clustering and the Search Optimization Service (SOS). The following image shows an example.

CHAPTER 8 DATA MODELING OPTIMIZATION STRATEGIES

```
SELECT start_time::DATE AS the_dt,
       schema_name,
       table_name,
       sum(rows_added) as row_added,
       sum(rows_updated) as rows_updated,
       sum(rows_removed) as rows_removed
FROM   snowflake.account_usage.table_dml_history
WHERE  start_time >= DATEADD(day, -7, CURRENT_TIMESTAMP())
AND schema_name='TPCH_SF1'
AND table_name='LINEITEM'
GROUP BY ALL
ORDER BY the_dt;
```

THE_DT	SCHEMA_NAME	TABLE_NAME	ROW_ADDED	ROWS_UPDATED	ROWS_REMOVED
2024-12-16	TPCH_SF1	LINEITEM	59986052	70660137	7329560
2024-12-19	TPCH_SF10	LINEITEM	0	179958156	0
2024-12-19	TPCH_SF10	ORDERS	0	15000000	0
2024-12-19	TPCH_SF1	LINEITEM_JSON_DEMO	119972124	0	59986052

After you understand the fundamentals and cost considerations of clustering, the next step is to evaluate the real-world impact. In the next section, we focus on how clustering can be applied to address specific performance challenges, particularly when large datasets are accessed through multiple query paths.

Cluster Key Usage

Query performance is the most reliable indicator of clustering effectiveness. When queries leverage clustering keys in local predicates, the PARTITIONS_SCANNED metric in the QUERY_HISTORY view offers a clear measure of efficiency. A lower percentage of partitions scanned relative to the total indicates better pruning and, therefore, greater benefit from clustering. To further evaluate clustering effectiveness, correlate this metric with clustering depth or clustering information for queries that utilize the clustering key.

Validate cluster key usage in micro-partition pruning from the query profile result:

```
--Validate Pruning
SELECT l_orderkey, l_extendedprice, l_quantity, o_orderstatus
FROM TPCH_SF1.LINEITEM
```

CHAPTER 8 DATA MODELING OPTIMIZATION STRATEGIES

```
JOIN TPCH_SF1.ORDERS
    ON l_orderkey = o_orderkey
WHERE l_shipdate = '2024-03-01';
```

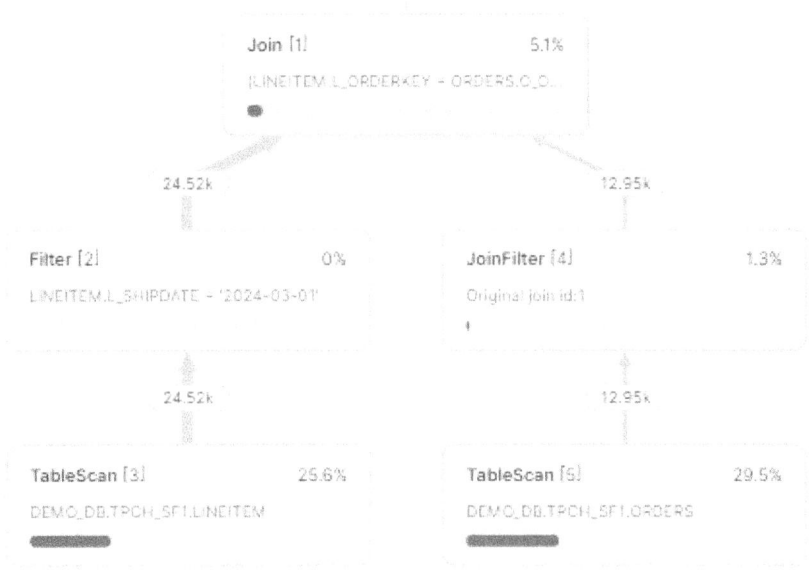

The query scanned a minimal subset of partitions relative to the total, demonstrating strong pruning performance.

Statistics

Scan progress	100.00%
Bytes scanned	10.13MB
Percentage scanned from ...	0.00%
Bytes written to result	0.04MB
Bytes sent over the netw...	8.94MB
Partitions scanned	13
Partitions total	95

Having explored how clustering can be applied to optimize query performance, the next step is to measure its actual impact. In the next section, we examine how to assess the effectiveness of clustering by analyzing pruning efficiency.

Determine Clustering Benefits

To evaluate pruning efficiency and assess how a table's default (natural) data ordering influences pruning effectiveness, utilize TABLE_PRUNING_HISTORY view. From the following query result, compare the number of pruned partitions to the total number of scanned partitions to determine the added value of the cluster key (the following image shows an example).

```
--Determine Clustering Benefits
SELECT
    table_name AS table_name,
    SUM(num_scans) AS total_num_scans,
    SUM(rows_scanned) AS total_rows_scanned,
    SUM(rows_pruned) AS total_rows_pruned,
    SUM(partitions_scanned) AS total_partitions_scanned,
    SUM(partitions_pruned) AS total_partitions_pruned,
    ((total_partitions_pruned / GREATEST(total_partitions_scanned , 1)) * 100)::decimal(12,2) AS pruning_ratio
  FROM SNOWFLAKE.ACCOUNT_USAGE.TABLE_PRUNING_HISTORY
  WHERE start_time >= DATEADD(day, -7, CURRENT_TIMESTAMP())
  GROUP BY ALL
  ORDER BY pruning_ratio DESC ;
```

	TABLE_NAME	TOTAL_NUM_SCANS	TOTAL_ROWS_SCANNED	TOTAL_ROWS_PRUNED	TOTAL_PARTITIONS_SCANNED	TOTAL_PARTITIONS_PRUNED
1	LINEITEM_JSON_DEMO	24	366453548	803253564	1170	2842
2	LINEITEM	36	1234051047	526987281	2064	770
3	ORDERS	13	84351696	16148304	221	41

Once the benefits of clustering have been evaluated through pruning metrics, it's equally important to understand the operational costs involved. The following section outlines how to monitor and interpret cost metrics associated with automatic clustering activities.

Cost Metrics for Clustering

The account usage view AUTOMATIC_CLUSTERING_HISTORY provides a history of automatic clustering operations, including the credits consumed, bytes updated, and rows updated each time a table is reclustered.

```
SELECT start_time::date as the_dt,
       schema_name,
       table_name,
       sum(credits_used) as credits_used,
       sum(num_bytes_reclustered) as num_bytes_reclustered
FROM snowflake.account_usage.AUTOMATIC_CLUSTERING_HISTORY
WHERE  start_time >= DATEADD(day, -7, CURRENT_TIMESTAMP())
AND schema_name='TPCH_SF1'
AND table_name='LINEITEM'
```

Moving on, we discuss materialized views and their benefits.

```
GROUP BY ALL
ORDER BY the_dt;
```

THE_DT	SCHEMA_NAME	TABLE_NAME	CREDITS_USED	NUM_BYTES_RECLUSTERED
2024-12-16	TPCH_SF1	LINEITEM	0.046950606	3412614144
2024-12-19	TPCH_SF1	LINEITEM_JSON_DEMO	0.065977611	2624487968

While clustering keys offers foundational performance improvements, they are just one part of a broader optimization toolkit. In the next section, we explore materialized views, an effective solution for accelerating complex or frequently executed queries by precomputing and storing their results.

Materialized Views

Another key feature of query acceleration is the materialized view. These views are designed to address a wide range of use cases, including performance bottlenecks that arise when accessing large datasets through multiple paths or subsets of table columns.

CHAPTER 8   DATA MODELING OPTIMIZATION STRATEGIES

In Snowflake, materialized views are a powerful mechanism for enhancing query performance, particularly for complex or frequently executed queries. By physically storing the results of these queries, materialized views eliminate the need for repeated execution of expensive calculations, resulting in significant performance gains. Furthermore, materialized views contribute to optimizing Snowflake data models in several key areas.

When to Create Materialized Views

- Performance improvement with frequent queries.
  - The query results contain a small number of rows and/or columns compared to the base table.
- The query results involve significant compute processing, such as:
  - Precompute complex aggregations and joins
    - Analysis of semi-structured data.
  - Aggregations that are time-consuming to calculate.
- Alternative clustering key for efficient pruning.
- Queries executed against external tables may exhibit slower performance compared to queries against native database tables or Apache Iceberg tables.
- Nonvolatile data in the base table.

Usage Optimization

This section examines the use of materialized views to address a specific performance challenge that arises when a large dataset has multiple access paths. Creating a materialized view in Snowflake allows you to specify a new clustering key, enabling Snowflake to reorganize the data during the initial creation of the materialized view.

CHAPTER 8   DATA MODELING OPTIMIZATION STRATEGIES

1) Create a table and load data to validate the materialized view:

```
USE DATABASE DEMO_DB;
CREATE SCHEMA TPCH_SF100;
USE SCHEMA TPCH_SF100;

CREATE OR REPLACE TABLE WEBLOG (
 CREATE_MS BIGINT,
 PAGE_ID BIGINT,
 TIME_TO_LOAD_MS INTEGER,
 METRIC2 INTEGER,
 METRIC3 INTEGER,
 METRIC4 INTEGER,
 METRIC5 INTEGER,
 METRIC6 INTEGER,
 METRIC7 INTEGER,
 METRIC8 INTEGER,
 METRIC9 INTEGER
);

INSERT INTO WEBLOG
SELECT
   (SEQ8())::BIGINT AS CREATE_MS
   ,UNIFORM(1,9999999,RANDOM(10002))::BIGINT PAGE_ID
   ,UNIFORM(1,9999999,RANDOM(10003))::INTEGER TIME_ON_LOAD_MS
   ,UNIFORM(1,9999999,RANDOM(10005))::INTEGER METRIC2
   ,UNIFORM(1,9999999,RANDOM(10006))::INTEGER METRIC3
   ,UNIFORM(1,9999999,RANDOM(10007))::INTEGER METRIC4
   ,UNIFORM(1,9999999,RANDOM(10008))::INTEGER METRIC5
   ,UNIFORM(1,9999999,RANDOM(10009))::INTEGER METRIC6
   ,UNIFORM(1,9999999,RANDOM(10010))::INTEGER METRIC7
   ,UNIFORM(1,9999999,RANDOM(10011))::INTEGER METRIC8
   ,UNIFORM(1,9999999,RANDOM(10012))::INTEGER METRIC9
 FROM TABLE(GENERATOR(ROWCOUNT => 1000000000))
 ORDER BY CREATE_ms;
```

CHAPTER 8   DATA MODELING OPTIMIZATION STRATEGIES

2) Verify clustering depth for an attribute on a table:
Effective clustering (see the following image):

```
SELECT SYSTEM$CLUSTERING_INFORMATION( 'WEBLOG' ,
'(CREATE_MS)' );

{ "cluster_by_keys" : "LINEAR(CREATE_MS)",
  "total_partition_count" : 1748,
  "total_constant_partition_count" : 0,
  "average_overlaps" : 0.0,
  "average_depth" : 1.0   .. }
```

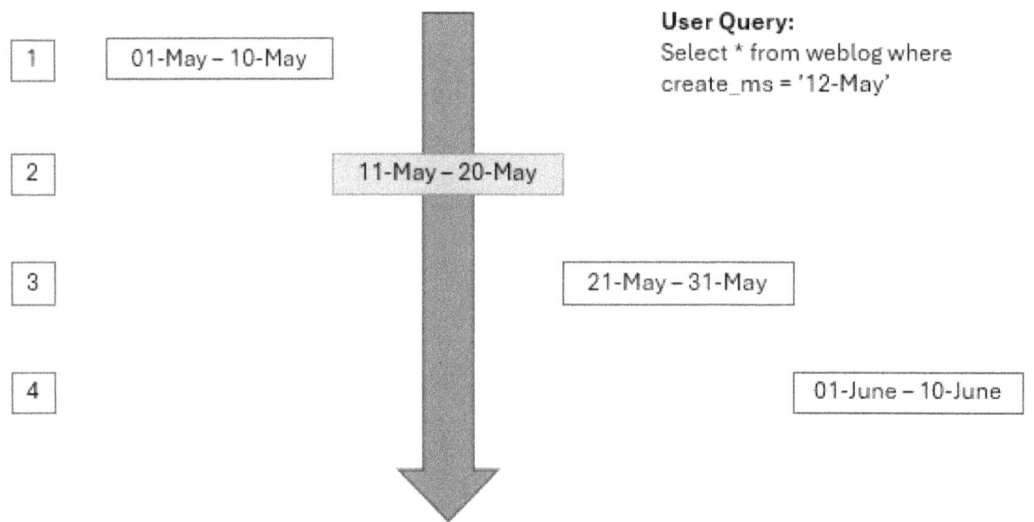

Ineffective clustering:

```
SELECT SYSTEM$CLUSTERING_INFORMATION( 'WEBLOG' , '(PAGE_ID)' );

{ "cluster_by_keys" : "LINEAR(PAGE_ID)",
  "total_partition_count" : 1748,
  "total_constant_partition_count" : 0,
  "average_overlaps" : 1747.0,
  "average_depth" : 1748.0  .. }
```

CHAPTER 8   DATA MODELING OPTIMIZATION STRATEGIES

Clustering depth for the CREATE_MS attribute is a good choice because it yields a low value—approximately 1 in this case. This indicates effective clustering. In contrast, the clustering depth for PAGE_ID is relatively high, which is suboptimal. However, this is expected since the data was generated in order of CREATE_MS, while PAGE_ID is randomly distributed.

3) Enable CREATE_MS as the cluster key:

   ALTER TABLE WEBLOG CLUSTER BY (CREATE_MS);

4) Validate performance using CREATE_MS:

   ALTER SESSION SET USE_CACHED_RESULT=FALSE;
   SELECT COUNT(*) CNT, AVG(TIME_TO_LOAD_MS) AVG_TIME_TO_LOAD
   FROM WEBLOG
   WHERE CREATE_MS BETWEEN 100000000 AND 100001000;

According to the Snowflake Query profile, the query completes in an average of 0.1 seconds, scanning only 1 out of 1,802 partitions (see the following image).

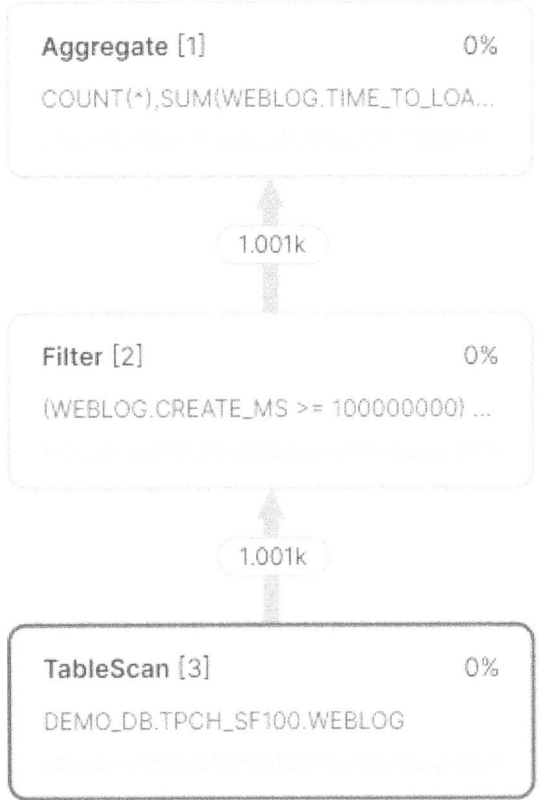

221

CHAPTER 8   DATA MODELING OPTIMIZATION STRATEGIES

5) Validate performance using PAGE_ID:

   ```
   ALTER SESSION SET USE_CACHED_RESULT=FALSE;
   SELECT COUNT(*) CNT, AVG(TIME_TO_LOAD_MS)  AVG_TIME_TO_LOAD
   FROM WEBLOG
   WHERE PAGE_ID=100000;
   ```

   According to the Snowflake Query profile, the query completes in an average of 100 seconds, scanning only 1801 out of 1802 partitions (see the following image).

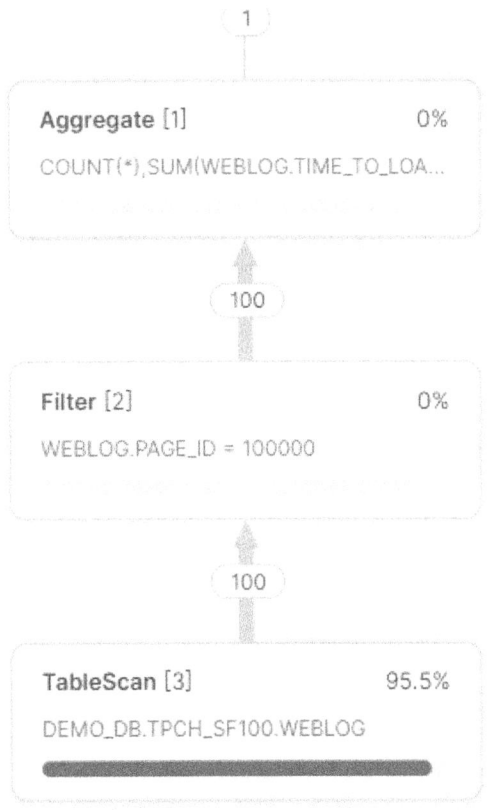

6) Create a clustered materialized view. The solution leverages Snowflake materialized views. By creating a materialized view with a new clustering key, Snowflake reorganizes the data during the initial creation.

   ```
   CREATE OR REPLACE MATERIALIZED VIEW MV_PAGE_ID
     (CREATE_MS, PAGE_ID, TIME_TO_LOAD_MS) CLUSTER BY (PAGE_ID)
   ```

CHAPTER 8 DATA MODELING OPTIMIZATION STRATEGIES

```
AS
SELECT CREATE_MS, PAGE_ID, TIME_TO_LOAD_MS
FROM WEBLOG;
```

Validate optimal data distribution by retrieving the clustering information

```
SELECT SYSTEM$CLUSTERING_INFORMATION( 'MV_PAGE_ID' ,
'(PAGE_ID)' );

{ "cluster_by_keys" : "LINEAR(PAGE_ID)",
  "total_partition_count" : 492,
  "total_constant_partition_count" : 0,
  "average_overlaps" : 1.9472,
  "average_depth" : 2.0 .. }
```

7) Validate performance on PAGE_ID where the optimizer uses the materialized view:

```
ALTER SESSION SET USE_CACHED_RESULT=FALSE;
SELECT COUNT(*),AVG(TIME_TO_LOAD_MS)
FROM WEBLOG
WHERE PAGE_ID=100000;
```

According to the Snowflake Query profile, the query completes in an average of 1 second, scanning only 1 out of 512 partitions from the materialized view (see the following image).

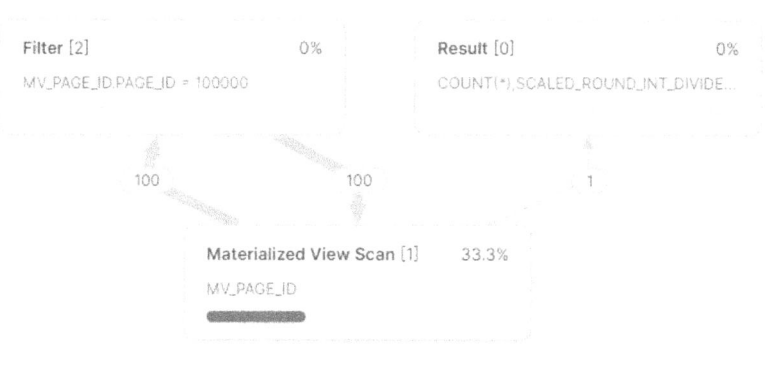

223

CHAPTER 8   DATA MODELING OPTIMIZATION STRATEGIES

Using materialized views effectively addresses a critical performance bottleneck when accessing data in a large table through different sets of attributes. The performance gains are substantial and eliminate the need for additional maintenance overhead, such as materialized view refresh jobs.

Usage Metrics for Materialized Views

Snowflake utilizes an operational virtual warehouse MATERIALIZED_VIEW_MAINTENANCE for materialized view maintenance activities. The credit consumption for these activities is recorded in the MATERIALIZED_VIEW_REFRESH_HISTORY view.

```
SELECT start_time::date AS the_dt,
  schema_name,
```

Next, we cover search optimization techniques.

```
  table_name,
  SUM(credits_used) AS credits_used
FROM snowflake.account_usage.materialized_view_refresh_history
WHERE start_time >= DATEADD(day, -7, CURRENT_TIMESTAMP())
GROUP BY ALL
ORDER BY 4 DESC;
```

THE_DT	SCHEMA_NAME	TABLE_NAME	CREDITS_USED
2024-12-21	TPCH_SF1	MV_LINEITEM_PART	0.019682358

While materialized views are ideal for accelerating complex queries and aggregations, they may not be the best fit for highly selective, point-lookup queries. In such cases, Search Optimization Service (SOS) offers a powerful alternative. The following section examines how this feature can substantially enhance performance for exact-match and selective filters, particularly on large datasets.

Search Optimization Service (SOS)

For those familiar with traditional table indexes, which enhance query performance for selective point-lookup queries on large tables, SOS in Snowflake offers similar capabilities. It is a powerful technique that accelerates the retrieval of exact matches for specific columns, making it an essential strategy for improving performance in large-scale data environments.

The following section outlines how search optimization can enhance Snowflake data modeling, focusing on several key guidelines and best practices.

- Ideal for point-lookup using specific columns on large tables
  - Equality searches
  - Substring searches
  - Semi-structured searches
  - Geospatial searches
- Ideal for non-clustered tables or queries not using clustering keys
- Recommended checks to consider:
  - Query filter operation has at least 100k-200k distinct values.
  - Query uses equality predicate or predicates that use IN.
  - Query returns a few rows with highly selective filters.
  - Query typically runs for at least tens of seconds.

Estimate Search Optimization Cost

The cost estimates provided by ESTIMATE_SEARCH_OPTIMIZATION are best-effort predictions of the cost for maintaining the SOS. Actual realized costs may vary by up to 50%, or in rare cases, by several times the estimated costs.

```
SELECT SYSTEM$ESTIMATE_SEARCH_OPTIMIZATION_COSTS('WEBLOG_SOS',
'EQUALITY(CREATE_MS,PAGE_ID,METRIC3)');
{
  "tableName" : "WEBLOG_SOS",
  "searchOptimizationEnabled" : true,
```

```
  "costPositions" : [ {
    "name" : "BuildCosts",
    "costs" : {
      "value" : 0.141207,
      "unit" : "Credits"
    },
    "computationMethod" : "Estimated",
    "comment" : "estimated via sampling"
  }, {
    "name" : "StorageCosts",
    "costs" : {
      "value" : 0.010259,
      "unit" : "TB",
      "perTimeUnit" : "MONTH"
    },
    "computationMethod" : "Estimated",
    "comment" : "estimated via sampling"
  }, {
    "name" : "Benefit",
    "computationMethod" : "NotAvailable",
    "comment" : "Currently not supported."
  }, {
    "name" : "MaintenanceCosts",
    "computationMethod" : "NotAvailable",
    "comment" : "Insufficient data to compute estimate for maintenance cost. Table is too young. Requires 7 day(s) of history."
  } ]
}
```

The summary of the cost estimate output is listed in Table 8-2.

CHAPTER 8 DATA MODELING OPTIMIZATION STRATEGIES

Table 8-2. *Search Optimization Cost Estimate*

Cost Represented	Description
Build costs	Estimated costs of building the search access path for the table. If search optimization has already been added to the table or to all specified columns, this object contains no cost information.
Storage costs	Estimated amount of storage space (in TB) required for the search access path for the table.
Maintenance costs	Estimated cost of maintaining the search access path for the table when rows are inserted, deleted, or modified. If the table has been created recently, no cost information is reported.

Usage Optimization

In this section, we examine how search optimization can be used to address a specific performance issue that arises during point-lookups.

```
CREATE TABLE WEBLOG_SOS AS SELECT * FROM WEBLOG;

SELECT approx_count_distinct(CREATE_MS),
       approx_count_distinct(PAGE_ID),
       approx_count_distinct(METRIC3)
FROM WEBLOG_SOS;
```

APPROX_COUNT_DISTINCT(CREATE_MS)	APPROX_COUNT_DISTINCT(PAGE_ID)	APPROX_COUNT_DISTINCT(METRIC3)
1017190496	9934432	9934432

Before Enabling SOS

Check query performance for different query access paths before enabling SOS.

```
select * from WEBLOG_SOS where CREATE_MS = 361690547;
-- Partitions Scanned: 234/1775   Time Taken: 2.0s   Records Returned: 1 row

select * from WEBLOG_SOS where PAGE_ID = 8710714;
-- Partitions Scanned: 1775/1775  Time Taken: 4.7s   Records Returned: 106 row
```

CHAPTER 8 DATA MODELING OPTIMIZATION STRATEGIES

```
select * from WEBLOG_SOS where METRIC3 = 3270447;
-- Partitions Scanned: 1775/1775   Time Taken: 4.5s   Records
Returned: 88 row
```

Enable SOS and Track Progress

Enable SOS and validate the performance for the same set of queries.

```
ALTER TABLE WEBLOG_SOS ADD SEARCH OPTIMIZATION ON EQUALITY(CREATE_MS,PAGE_
ID,METRIC3);
```

Track Search Optimization Progress

```
SHOW TABLES LIKE '%WEBLOG_SOS%';
SELECT "schema_name",
       "name" as table_name,
       "search_optimization",
       "search_optimization_progress",
       "search_optimization_bytes"
FROM table(result_scan(last_query_id()));
```

schema_nam	TABLE_NAME	search_optimization	search_optimization_progress	search_optimization_bytes
TPCH_SF100	WEBLOG_SOS	ON	100	11302366208

After Enabling SOS

Execute the same set of queries and measure the performance.

```
select * from WEBLOG_SOS where CREATE_MS = 361690547;
-- Partitions Scanned: 1/1775   Time Taken: 33ms   Records Returned: 1 row

select * from WEBLOG_SOS where PAGE_ID = 8710714;
-- Partitions Scanned: 105/1775   Time Taken: 1.6s   Records
Returned: 106 row
```

```
select * from WEBLOG_SOS where METRIC3 = 3270447;
-- Partitions Scanned: 87/1775   Time Taken: 1.3s   Records Returned: 88 row
```

Comparison Metrics

Before enabling SOS, queries on the WEBLOG_SOS table exhibited high latency and scanned a large number of partitions, leading to inefficient performance. After enabling SOS, the same queries showed significant improvements in both speed and resource utilization. Table 8-3 presents a comparison of query performance before and after enabling SOS.

Table 8-3. Query Performance Before and After Enabling SOS

Query	Partitions Scanned	Time Taken	Improvement After Enabling SOS
Query-1	234 → 1	2.0s → 33ms	Drastic reduction in partitions scanned and execution time
Query-2	1775 → 105	4.7s → 1.6s	Significant improvement in scan efficiency and speed
Query-3	1775 → 87	4.5s → 1.3s	Substantial reduction in scan scope and latency

Confirm the use of SOS by reviewing the Query Profiler results (shown here), which should show that SOS lead to optimized access paths and reduced partition scans.

CHAPTER 8   DATA MODELING OPTIMIZATION STRATEGIES

Determine Search Optimization Benefits

To assess the effectiveness of pruning resulting from search optimization, refer to the SEARCH_OPTIMIZATION_BENEFITS view. This view provides insights into the metrics outlined in Table 8-4.

CHAPTER 8 DATA MODELING OPTIMIZATION STRATEGIES

Table 8-4. *Search Optimization Benefits Metrics*

Cost Represented	Description
PARTITIONS_SCANNED	The number of partitions scanned during the scan operations described in NUM_SCANS.
PARTITIONS_PRUNED_DEFAULT	The number of partitions pruned as a result of the default (natural) ordering of data during the scan operations described in NUM_SCANS.
PARTITIONS_PRUNED_ADDITIONAL	The number of partitions pruned as a result of search optimization during the scan operations described in NUM_SCANS.

```
SELECT
    start_time::date AS the_dt,
    schema_name,
    table_name,
    SUM(num_scans) AS num_scans,
    SUM(partitions_pruned_default) AS partitions_pruned_default,
    SUM(partitions_pruned_additional) AS partitions_pruned_additional,
    SUM(partitions_scanned) AS partitions_scanned
FROM SNOWFLAKE.ACCOUNT_USAGE.SEARCH_OPTIMIZATION_BENEFITS
WHERE start_time >= DATEADD(day, -7, CURRENT_TIMESTAMP())
GROUP BY ALL
ORDER BY partitions_pruned_additional DESC;
```

THE_DT	SCHEMA_NAME	TABLE_NAME	NUM_SCANS	PARTITIONS_PRUNED_DEFAULT	PARTITIONS_PRUNED_ADDITIONAL	PARTITIONS_SCANNED
2024-12-21	TPCH_SF1	LINEITEM	3	0	16	972

Usage Metrics for Search Optimization

The credit usage associated with building and maintaining search optimization is recorded in SEARCH_OPTIMIZATION_HISTORY view.

```
SELECT start_time::date AS the_dt,
  schema_name,
  table_name,
```

Let's explore hybrid tables and their advantages for mixed workloads.

```
  SUM(credits_used) AS credits_used
FROM SNOWFLAKE.ACCOUNT_USAGE.SEARCH_OPTIMIZATION_HISTORY
WHERE start_time >= DATEADD(day, -7, CURRENT_TIMESTAMP())
GROUP BY ALL
ORDER BY 4 DESC;
```

THE_DT	SCHEMA_NAME	TABLE_NAME	CREDITS_USED
1 2025-01-05	TPCH_SF100	SEARCH OPTIMIZATION ON TABLE_ID:13322	0.140677546
2 2025-01-05	TPCH_SF100	SEARCH OPTIMIZATION ON TABLE_ID:5126	0.029914550

In addition to query acceleration techniques, Snowflake also supports advanced table types designed for specific workload patterns. One such capability is *hybrid tables,* which are purpose-built to handle mixed workloads that demand both high-throughput transactional processing and fast analytical performance.

Hybrid Tables for Mixed Workloads

When dealing with mixed workloads that combine operational (OLTP) and analytical (OLAP) queries—particularly those requiring low latency, high throughput, and efficient handling of small, random point reads and writes—consider utilizing hybrid tables.

> **Note** As of the time of writing, the Hybrid Tables feature is available only in commercial AWS regions. It is not yet supported on Azure or GCP platforms.

Hybrid tables utilize a row store as their underlying data storage, making them well-suited for achieving strong operational query performance. When data is written to a hybrid table, it is directly inserted into the row store. Subsequently, the data is asynchronously extracted to object storage. This approach enhances performance, improves workload isolation for large scans, and minimizes disruption to ongoing operational workloads. For optimal performance on analytical queries, some data may be cached in columnar format within your warehouse.

CHAPTER 8 DATA MODELING OPTIMIZATION STRATEGIES

Since hybrid tables have a row store as the primary storage, they typically exhibit a larger storage footprint compared to standard tables. The primary contributor to the difference is that columnar data for regular tables is frequently able to realize much higher compression ratios.

Hybrid Table Cost Components

- **Hybrid table storage:** Storage is charged at a flat monthly cost per gigabyte (GB) stored in hybrid tables.

- **Virtual warehouse compute:** Queries against hybrid tables execute through virtual warehouses. The warehouse consumption rate for querying hybrid tables is equivalent to standard tables.

- **Serverless usage:** Hybrid tables leverage serverless resources within the underlying row storage clusters. As a result, they incur additional credits compared to standard tables. Consumption is calculated according to the volume of data read from or written to these clusters. Credits are also spent on the compute resources that are consumed to perform background operations like data compaction.

Usage Optimization

In this section, we focus on using hybrid tables to address a specific performance challenge that arises when a point-lookup is required.

1) Create a hybrid table:

```
CREATE OR REPLACE HYBRID TABLE CUSTOMER_HYD_TBL (
    C_CUSTKEY NUMBER(38,0) PRIMARY KEY,
    C_NAME VARCHAR(25),
    C_ADDRESS VARCHAR(40),
    C_NATIONKEY NUMBER(38,0),
    C_PHONE VARCHAR(15),
    C_ACCTBAL NUMBER(12,2),
    C_MKTSEGMENT VARCHAR(10),
    C_COMMENT VARCHAR(117)
);
```

CHAPTER 8   DATA MODELING OPTIMIZATION STRATEGIES

2) Load sample data into the hybrid table:

   ```
   INSERT INTO CUSTOMER_HYD_TBL
   SELECT * FROM TPCH_SF10.CUSTOMER LIMIT 10000;
   ```

3) Create an index on the hybrid table:

   ```
   CREATE OR REPLACE INDEX idx_nation_key ON CUSTOMER_HYD_TBL
   (C_NATIONKEY);
   ```

   ```
   SHOW INDEXES;
   ```

   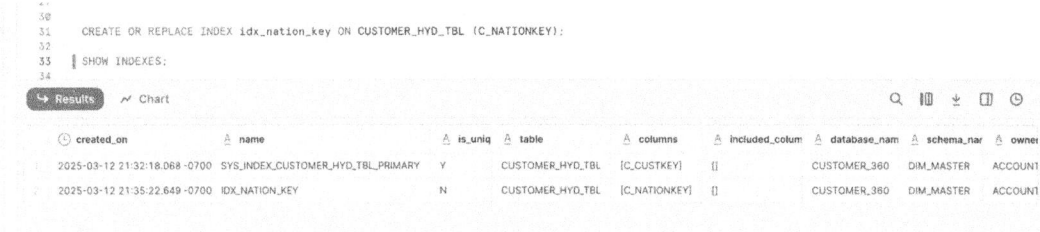

4) Perform DML activity on the hybrid table:

   ```
   DELETE CUSTOMER_HYD_TBL WHERE C_NATIONKEY=22;

   UPDATE CUSTOMER_HYD_TBL SET C_COMMENT = UPPER(C_COMMENT);

   ALTER SESSION SET QUERY_TAG = 'Add New Customer';
   INSERT CUSTOMER_HYD_TBL VALUES (325001,'John','J.F.K Street',1,'
   1-352-675-4934',8207.46,'FURNITURE','express, final braids');
   INSERT CUSTOMER_HYD_TBL VALUES (325002,'John','J.F.K Street',1,'
   1-352-675-4934',8207.46,'FURNITURE','express, final braids');
   INSERT CUSTOMER_HYD_TBL VALUES (325003,'John','J.F.K Street',1,'
   1-352-675-4934',8207.46,'FURNITURE','express, final braids');
   INSERT CUSTOMER_HYD_TBL VALUES (325004,'John','J.F.K Street',1,'
   1-352-675-4934',8207.46,'FURNITURE','express, final braids');
   INSERT CUSTOMER_HYD_TBL VALUES (325005,'John','J.F.K Street',1,'
   1-352-675-4934',8207.46,'FURNITURE','express, final braids');
   ALTER SESSION UNSET QUERY_TAG;
   ```

CHAPTER 8 DATA MODELING OPTIMIZATION STRATEGIES

Usage Metrics for Hybrid Tables

The credit usage involved in building and maintaining search optimization is recorded in the AGGREGATE_QUERY_HISTORY view.

1) Monitor the hybrid tables' workload:

```
SELECT
      query_tag
    , SUM(total_elapsed_time:"sum"::NUMBER) / SUM (calls) as avg_latency
    , SUM(hybrid_table_requests_throttled_count) as hybrid_table_requests_throttled_count
    , any_value(query_text)
FROM snowflake.account_usage.aggregate_query_history
WHERE query_tag ILIKE '%Add New Customer%'
AND interval_start_time >= DATEADD(day, -7, CURRENT_TIMESTAMP())
GROUP BY ALL
ORDER BY avg_latency DESC;
```

QUERY_TAG	AVG_LATENCY	HYBRID_TABLE_REQUESTS_THROTTLED_COUNT	ANY_VALUE(QUERY_TEXT)
Add New Customerer	220.222222	0	INSERT INTO CUSTOMER_HYD_TBL VALUES (325002,'John','J.F.K Street',1,'1-352-675-4934',82

2) Monitor storage usage by the hybrid tables:

```
SELECT USAGE_DATE,
       SUM(HYBRID_TABLE_STORAGE_BYTES)/ POWER(1024, 2) AS HYBRID_TABLE_STORAGE_MB
FROM SNOWFLAKE.ACCOUNT_USAGE.STORAGE_USAGE
WHERE USAGE_DATE >= DATEADD(day, -7, CURRENT_DATE())
GROUP BY ALL
ORDER BY 1;
```

CHAPTER 8   DATA MODELING OPTIMIZATION STRATEGIES

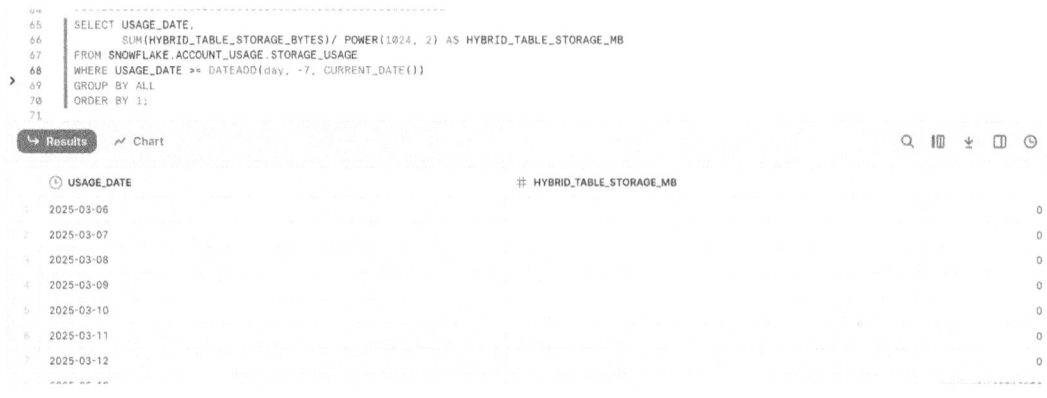

3) Monitor credit usage by the hybrid tables:

```
SELECT start_time::date as the_date,
   SUM(credits_used) AS total_credits
FROM SNOWFLAKE.ACCOUNT_USAGE.HYBRID_TABLE_USAGE_HISTORY
WHERE start_time >= DATEADD(day, -7, CURRENT_TIMESTAMP())
GROUP BY ALL
ORDER BY 1;
```

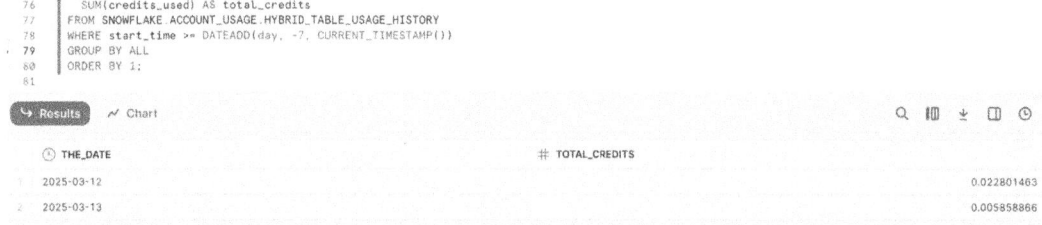

In addition to optimizing for mixed workloads, Snowflake also supports techniques to streamline joins in complex data models. One such method is soft referential integrity, which helps improve performance by guiding the optimizer to skip join enforcement.

Establishing Soft Referential Integrity

Soft referential integrity allows Snowflake to optimize query performance by skipping unnecessary joins when foreign key relationships are declared using the RELY keyword. This technique provides the optimizer with valuable relationship context without enforcing constraints at the database level, making it ideal for complex data models and enhancing query efficiency.

Skip Unnecessary Joins

In a typical data model, fact tables are linked to dimension tables through foreign key relationships. During query execution, Snowflake's optimizer must determine the necessity of each join to retrieve the requested data. Without a clear indication of the relationships between fact and dimension tables, the optimizer may perform unnecessary joins, potentially impacting query performance negatively.

By establishing soft referential integrity using foreign key constraints with the RELY keyword, you provide the optimizer with valuable hints about the relationships between tables. The RELY keyword signals to Snowflake that these foreign key constraints are not strictly enforced (soft integrity), but are intended as informational cues for query optimization. This enables Snowflake to avoid unnecessary joins during query execution, particularly when the columns in the foreign key relationship are not referenced in the query. As a result, significant performance improvements can be achieved by eliminating unnecessary joins, which often involve processing large volumes of data.

Eliminate Unneeded Data Scans

Beyond eliminating unnecessary joins, soft referential integrity contributes to reducing the volume of data scanned during query execution. By leveraging foreign key constraints, Snowflake's optimizer can effectively identify the tables likely to be relevant for a query, even when joins are not explicitly specified in the query itself.

For instance, if a query filters data based on a column within the fact table but does not reference any columns from the associated dimension table, the optimizer can avoid scanning the dimension table entirely. This pruning of unnecessary data can significantly reduce the I/O operations required to fulfill the query, improving both speed and cost efficiency, especially when working with large datasets.

CHAPTER 8 DATA MODELING OPTIMIZATION STRATEGIES

Usage Optimization

This section explores how soft referential integrity (RELY) can be utilized to address a specific performance challenge that arises when joining tables with referential relationships.

1) Prepare the sample data:

```
USE SCHEMA DEMO_SCHM;
USE SCHEMA TPCH_SF10;
CREATE TABLE ORDERS AS
    SELECT * FROM SNOWFLAKE_SAMPLE_DATA.TPCH_SF100.ORDERS;
CREATE TABLE LINEITEM AS
    SELECT * FROM SNOWFLAKE_SAMPLE_DATA.TPCH_SF100.LINEITEM;
```

2) Create a view joining both tables:

```
CREATE VIEW TPCH_SF10.v_lineitem_orders
AS
SELECT l_orderkey, l_extendedprice, l_quantity, o_orderstatus
FROM TPCH_SF10.LINEITEM
JOIN TPCH_SF10.ORDERS
    ON l_orderkey = o_orderkey;
```

Validate join elimination when there is no reference to the [orders] table. For example, the following query on v_lineitem_orders has no reference to any columns from the [orders] table. When this query is executed, Snowflake does not automatically eliminate the join.

```
SELECT  l_orderkey, sum(l_extendedprice*l_quantity),
current_timestamp()
FROM TPCH_SF10.v_lineitem_orders
GROUP BY ALL;
```

CHAPTER 8 DATA MODELING OPTIMIZATION STRATEGIES

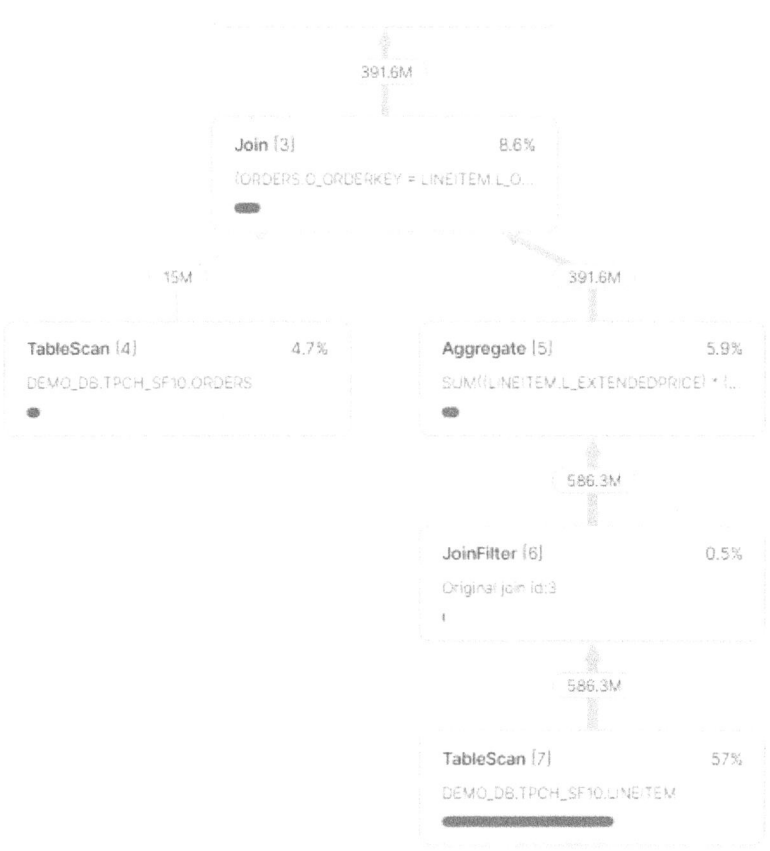

3) Enable the RELY key reference:

   ```
   ALTER TABLE TPCH_SF10.ORDERS
        ADD CONSTRAINT pk_orderkey PRIMARY KEY ( o_orderkey ) RELY;
   ALTER TABLE TPCH_SF10.LINEITEM
        ADD CONSTRAINT fk_lineitem_order FOREIGN KEY ( l_orderkey )
        REFERENCES TPCH_SF10.ORDERS ( o_orderkey ) RELY;
   ```

4) Check the table constraints:

   ```
   SELECT constraint_name, table_name, constraint_type, enforced
   FROM information_schema.table_constraints
   ORDER BY created ASC;
   ```

CHAPTER 8   DATA MODELING OPTIMIZATION STRATEGIES

CONSTRAINT_NAME	TABLE_NAME	CONSTRAINT_TYPE	ENFORCED
PK_ORDERKEY	ORDERS	PRIMARY KEY	NO
FK_LINEITEM_ORDER	LINEITEM	FOREIGN KEY	NO

5) Validate join elimination using SoftRI:

```
SELECT   l_orderkey, sum(l_extendedprice*l_quantity), current_timestamp()
FROM TPCH_SF10.v_lineitem_orders
GROUP BY ALL;
```

Confirm the use of Soft-RI, which eliminates the join on the [orders] table by reviewing the Query Profiler results, as shown here.

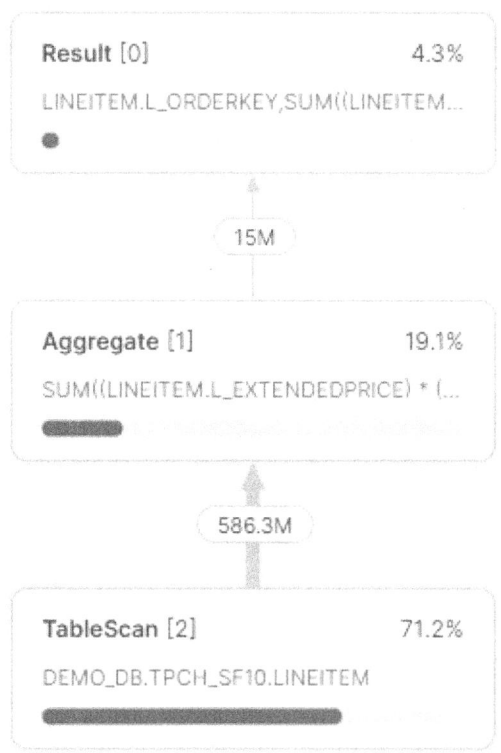

> **Note** Join elimination could produce incorrect results if the constraint's `RELY` property is used and the data does not comply with the `UNIQUE`, `PRIMARY KEY`, and `FOREIGN KEY` constraints. Hence, it is important to understand the implications of this optimization.

In addition to optimizing structured data models, Snowflake also provides robust capabilities for working with semi-structured data. The next section explores how to efficiently store, query, and optimize formats like JSON using the `VARIANT` datatype.

Handling Semi-Structured Data

Initially, when working with JSON data in Snowflake, its ease of use and scalability were evident advantages. However, upon closer examination, we recognized the potential for high credit consumption when dealing with large datasets. To mitigate this challenge, we restructured our tables and optimized our JSON data storage strategy. This resulted in a significant reduction in bytes scanned, disk spillage, and query execution time.

Advanced Metadata Engine

The `VARIANT` datatype in Snowflake is a versatile, tagged format that can hold up to 16 MB of any type of semi-structured data. Snowflake distinguishes itself by extending beyond the traditional database approach, simplifying the storage and access of semi-structured data with a flexible schema. This flexibility empowers users to work with data in ways that were previously far more complex.

The `VARIANT` datatype optimizes JSON query performance by extracting many elements (around 200 elements) in columnar form and building an *advanced metadata engine*. Snowflake identifies and stores frequently repeated attributes across records separately, leading to improved compression and faster data access. This approach mirrors the columnar storage principles employed by traditional databases, as shown in the following image.

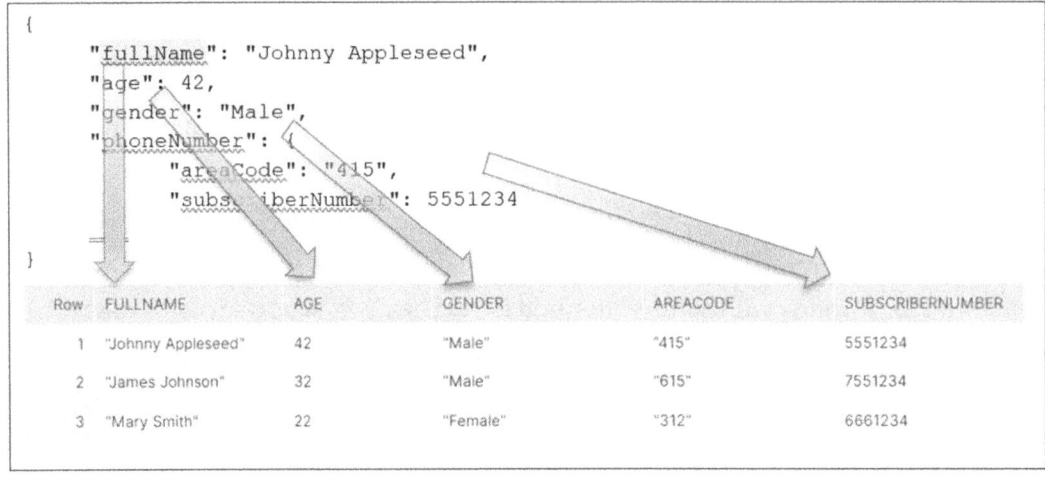

In addition, Snowflake collects, calculates, and stores statistics about the sub-columns within the metadata repository. This gives the Snowflake advanced query optimizer metadata about the semi-structured data to optimize the data access. The collected statistics allow the optimizer to use pruning to minimize the amount of data accessed, thereby accelerating data retrieval.

There are some limitations regarding the extraction of elements with certain characteristics into columns, for example:

- Any elements containing null are not extracted

- An element could have different datatypes that are not extracted, i.e., {"foo":1} and {"foo":"1"}

- Elements with missing values are not extracted

How JSON Extraction Impacts Queries

When it comes to performance, querying JSON data versus a traditional relational table may produce similar results in some cases; in other cases, query performance against JSON is not as optimal as on a relational table. Here's why:

- Queries on semi-structured data do not use the result cache.

- In semi-structured formats, each JSON document is treated as a single column, whereas relational tables store one element per column.

- When querying a semi-structured element, Snowflake's execution engine behaves differently depending on whether the element was extracted:

  - If the element was extracted into an advanced metadata column, the engine scans only the extracted column.

  - If the element was not extracted into an advanced metadata column, the engine scans the entire JSON structure and traverses it for each row to output values, which can negatively impact performance.

- JSON NULL differs from SQL NULL. JSON NULL represents element values that are assigned the string 'null', which prevents these elements from being extracted. SQL NULL represents an empty value. As a result, JSON values with a 'null' can have a negative impact on query performance.

Tips for Optimizing Query Performance on JSON Data

When designing a table that stores JSON data in the VARIANT column, follow these tips:

- Flatten frequently accessed attributes from JSON before loading the data, while still storing the full JSON structure in the VARIANT column.

- If the "null" values in your files simply represent missing data and don't have any special meaning, set the STRIP_NULL_VALUES file format option to TRUE when loading semi-structured data. This will convert the JSON NULL values into SQL NULL values.

- Ensure that each unique element in your data is stored using a single, native datatype (e.g., string or number for JSON).

- Avoid creating tables with an excessive number of columns (e.g., thousands). Tables with tens or hundreds of columns typically exhibit better performance.

- It's generally better not to mix very wide columns (like VARIANT) with very small datatypes.

For example, if a query is accessing only the smaller columns, performance may suffer because it has to scan a larger block that includes the wider VARIANT columns. A good solution is to create two separate tables: one for non-variant datatype columns and another for VARIANT columns, maintaining a one-to-one relationship on the keys for easy joins.

Usage Optimization

This section explores effective use cases of JSON extraction, focusing on the importance of leveraging the native JSON datatype for applying predicate conditions to prevent performance degradation. We begin by examining methods to identify the datatypes of attributes within JSON data, a crucial step for efficiently querying and processing.

Find the JSON Attribute Datatype

Use the TYPEOF function to find the datatype of a specific JSON attribute or the datatype of an entire JSON object. This function returns the datatype attributes—the integer, string, array, Boolean, or another JSON datatype—which is useful when dealing with complex, nested JSON structures.

```
SELECT
    typeof(l_payload) as payload_type,
    l_payload:l_orderkey as l_orderkey,
    l_payload:l_shipdate as l_shipdate,
    typeof(l_payload:l_orderkey) as l_orderkey_dtype,
    typeof(l_payload:l_shipdate) as l_shipdate_dtype,
FROM TPCH_SF1.LINEITEM_JSON_DEMO  LIMIT 10;
```

	PAYLOAD_TYPE	L_ORDERKEY	L_SHIPDATE	L_QUANTITY	O_ORDERDATE	O_TOTALPRICE
1	OBJECT	INTEGER	DATE	DECIMAL	DATE	DECIMAL
2	OBJECT	INTEGER	DATE	DECIMAL	DATE	DECIMAL
3	OBJECT	INTEGER	DATE	DECIMAL	DATE	DECIMAL
4	OBJECT	INTEGER	DATE	DECIMAL	DATE	DECIMAL
5	OBJECT	INTEGER	DATE	DECIMAL	DATE	DECIMAL
6	OBJECT	INTEGER	DATE	DECIMAL	DATE	DECIMAL

Test Case: Optimize Filter Conditions Using Physical Columns

Use physical (non-derived) columns in filter conditions to enable effective data pruning. This approach helps Snowflake skip unnecessary micro-partitions, significantly improving query performance and reducing compute costs.

```
ALTER SESSION SET USE_CACHED_RESULT = FALSE ;
ALTER SESSION SET QUERY_TAG='UseCase-1: Use Physical Column in
Filter Cont.';
SELECT    l_orderkey,
          l_shipdate,
          l_payload:l_orderkey as l_orderkey,
          l_payload:l_shipdate as l_shipdate,
          l_payload:l_quantity as l_quantity,
          l_payload:o_orders[0].o_orderdate as o_orderdate,
          l_payload:o_orders[0].o_totalprice
FROM TPCH_SF1.LINEITEM_JSON_DEMO
WHERE l_shipdate BETWEEN '2022-01-01' and '2022-01-31';
```

Confirm effective partition pruning by examining the Query Profiler for minimized data scan indicators, as shown in the following image.

Statistics

Scan progress	100.00%
Bytes scanned	11.63MB
Percentage scanned from ...	0.00%
Bytes written to result	13.06MB
Bytes sent over the netw...	0.06MB
Partitions scanned	3
Partitions total	1617

Measure Performance Metrics

For detailed information on partition pruning, fetch query execution metrics from QUERY_HISTORY.

```
SELECT  QUERY_TAG,
        TOTAL_ELAPSED_TIME,
        BYTES_SCANNED,
        PARTITIONS_TOTAL,
        PARTITIONS_SCANNED,
        ROWS_PRODUCED
FROM SNOWFLAKE.ACCOUNT_USAGE.QUERY_HISTORY
WHERE QUERY_TAG iLike ('%UseCase%')
AND QUERY_TYPE = 'SELECT'
ORDER BY start_time;
```

QUERY_TAG	TOTAL_ELAPSED_TIME	BYTES_SCANNED	PARTITIONS_TOTAL	PARTITIONS_SCANNED	ROWS_PRODUCED
UseCase-1: Use Physical Column in Filter Cont.	6468.0	12091392	1617	3	772005
UseCase-2: Use JSON attribute; Match Datatype	14810.5	205004800	1617	1617	772005
UseCase-1: Use Physical Column in Filter Cont.	4843.0	12189696	1617	3	772005
UseCase-2: Use JSON attribute; Match Datatype	14204.4	242442240	1617	1617	772005

There are many benefits to denormalizing fact and dimension tables.

Semi-structured Data Dos and Don'ts

Follow these guidelines to ensure efficient querying and optimal performance when working with semi-structured JSON data.

- Use physical columns in predicate conditions for optimal performance.
- Do not use PARSE_JSON unless necessary.
- Do not use datatype conversion functions. Instead, try to match the type defined in the JSON.

While effective handling of semi-structured data lays a strong foundation for performance, further optimization can be achieved through denormalization. Denormalization is a technique that reduces query complexity and accelerates execution in Snowflake environments.

Denormalizing Fact/Dimension Tables

By denormalizing fact/dimension relationships and incorporating dimension attributes directly into the fact table, Snowflake can significantly reduce the computational overhead associated with traditional normalized schemas. However, the benefits of this approach must be carefully evaluated to ensure its suitability for the specific use case.

Why Denormalize Fact/Dim Tables?

Snowflake's query optimization ensures that queries are processed efficiently, but frequent joins between fact and dimension tables can still degrade performance. Fact/Dim models often rely on views containing multiple joins to combine data, but Snowflake executes these joins for every query, even if columns from the associated tables are not selected. As a result, queries can become slower, particularly when dealing with large datasets or complex relationships.

Denormalizing tables by storing dimension descriptions directly within the fact table removes the need for Snowflake to perform unnecessary joins. This reduction in joins speeds up query performance, as fewer tables need to be accessed, and the optimizer can easily prune partitions based on the fact data. The result is improved performance, lower resource consumption, and reduced compute costs, making the process much more efficient.

Addressing Diminishing Returns

Denormalization offers clear performance benefits, but it's important to be aware of diminishing returns when you start adding too many columns. For instance, adding hundreds or thousands of fields to a table can increase I/O, especially in smaller-sized warehouses. As more partitions are created and queried, this leads to additional disk reads, which can offset the performance improvements you were aiming for.

Moreover, large numbers of columns can lead to inefficient data compression. Fields that do not compress efficiently—such as large text fields (`VARCHAR`)—could negatively impact storage and query performance. This is why denormalization should be done judiciously, and the "break-even point" for the number of columns or data size must be evaluated on a case-by-case basis.

For organizations unable to denormalize fully, alternative strategies such as soft referential integrity and combining dimensional tables can provide a middle ground for improved performance without the full overhead of denormalization.

Summary

This chapter presented essential strategies for optimizing data models in Snowflake to boost performance, scalability, and cost efficiency. It emphasized aligning data structures with query patterns using techniques like clustering keys for pruning, materialized views for precomputed results, and search optimization for selective lookups. The chapter also explored hybrid tables for mixed workloads and soft referential integrity to reduce unnecessary joins. For semi-structured data, we recommend leveraging the VARIANT type and flattening JSON attributes. Finally, this chapter highlighted the benefits and tradeoffs of denormalizing fact and dimension tables to minimize joins and improve query speed.

With an optimized data model in place, the next step is to ensure that data loading processes are equally efficient. Optimizing how data is ingested into Snowflake can significantly impact overall system performance and resource utilization.

CHAPTER 9

Data Load Optimization Strategies

Overview

Data loading is a foundational step in any data platform. In Snowflake, loading data efficiently and reliably is critical to enabling downstream analytics, reporting, and machine learning. Whether we are ingesting structured files, semi-structured logs, or streaming data, Snowflake offers a flexible and powerful set of tools to support diverse data loading needs.

This chapter provides a practical guide to Snowflake data loading, including planning strategies, use cases, benefits, and considerations to help you design robust and scalable ingestion pipelines.

The essential data load optimization strategies covered in this chapter include:

- Using `COPY INTO` with optimal file sizes (100–250 MB compressed) to balance parallelism across all compute nodes within a warehouse.

- Leveraging Snowpipe for continuous, near real-time ingestion with event-based triggers and serverless architecture.

- Adopting external tables and Apache Iceberg to reduce data duplication and enable multi-engine access without reloading data.

- Choosing efficient file formats like Parquet or ORC to improve compression, reduce I/O, and enhance query performance.

- Monitoring and tuning load performance using Snowflake Query Profile and Account Usage views to identify bottlenecks and optimize resource usage.

CHAPTER 9   DATA LOAD OPTIMIZATION STRATEGIES

To build efficient and scalable data pipelines in Snowflake, it is essential to go beyond basic ingestion techniques and adopt practices that enhance performance, reduce costs, and simplify operations. The strategies covered in this chapter provide a practical framework for optimizing data loads across various use cases and data sources.

Data Load Optimization Strategies

Efficient data loading is the key to performance and cost savings in Snowflake. Whether loading from cloud storage or streaming sources, a well-optimized process matters. It improves query speed through well-organized micro-partitioning. It also reduces compute and storage usage. This leads to more efficient cost management. Figure 9-1 highlights the main strategies covered in this section.

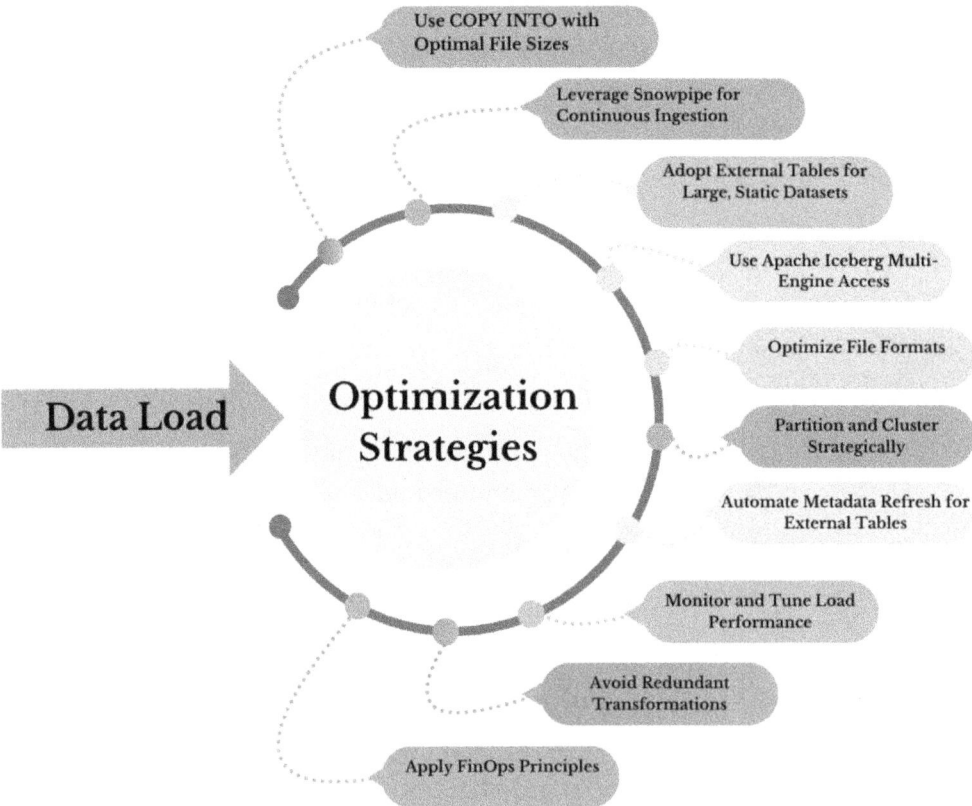

Figure 9-1. Data load optimization strategies

Use COPY INTO with Optimal File Sizes

When loading data from cloud storage (e.g., Amazon S3, Azure Blob, or GCS), Snowflake recommends using the COPY INTO command with files sized between 100 MB and 250 MB compressed. This size range balances parallelism and metadata overhead.

Best practices:

- Split large files using tools like AWS Glue or Apache Spark.
- Avoid loading many small files, which can increase metadata and processing time.

Leverage Snowpipe for Continuous Ingestion

For near real-time data loading, Snowpipe offers an automated and scalable solution. It loads data as soon as it lands in cloud storage, reducing latency and manual intervention.

Best practices:

- Use event-based triggers (e.g., AWS S3 event notifications).
- Monitor Snowpipe usage to avoid unnecessary invocations and costs.

Adopt External Tables for Large, Static Datasets

Instead of loading large datasets into Snowflake, consider using external tables to query data directly from cloud storage. This is ideal for historical or infrequently accessed data.

Best practices:

- Organize data in partitioned directories.
- Use partition columns in queries to enable pruning and reduce scan costs.

Note For detailed guidance on using external tables, refer to the section entitled "Manage the Storage Lifecycle" in Chapter 7. It outlines the steps for adopting external tables effectively.

CHAPTER 9   DATA LOAD OPTIMIZATION STRATEGIES

Use Apache Iceberg for Multi-Engine Access

Apache Iceberg tables allow Snowflake and other engines (like BigQuery or DuckDB) to access the same data without duplication. This is especially useful for reducing ETL redundancy.

Benefits:

- One copy of the data for all tools.
- Reduced storage and compute costs.
- Supports schema evolution and Time Travel.

Optimize File Formats

Choose efficient file formats like Parquet or ORC over JSON or CSV. These columnar formats reduce storage and improve query performance.

Why it matters:

- Columnar formats support predicate pushdown and compression.
- Columnar formats reduce I/O and improve scan efficiency.

Partition and Cluster Strategically

Use the Query Profile and Account Usage views to monitor load performance, identify bottlenecks, and analyze query behavior.

Best practices:

- Partition external tables by date or region using directory-based storage. This allows early pruning of unnecessary files during query planning, improving performance.
- Use clustering keys on large internal tables to improve performance over time.

Automate Metadata Refresh for External Tables

For externally managed Iceberg tables, Snowflake requires periodic metadata refreshes to stay in sync with new data.

How to do this:

- Create a stored procedure to check and refresh metadata.
- Schedule it using Snowflake Tasks.

Monitor and Tune Load Performance

Use the Query Profile and Account Usage views to monitor load performance and identify bottlenecks.

Metrics to watch:

- Total time taken to load data
- File size distribution; imbalanced sizes can affect parallelism
- Errors and retries

Avoid Redundant Transformations

Minimize unnecessary data transformations during load. Instead, use tools like dbt to handle transformations after data is loaded.

Why it matters:

- It reduces compute usage during load.
- It keeps ETL pipelines simpler and more maintainable.

Apply FinOps Principles

Track and optimize data load costs using FinOps practices to ensure efficient resource usage and cost transparency. FinOps brings together engineering, finance, and business teams to collaborate on cloud spending decisions, enabling smarter resource usage and budget alignment.

CHAPTER 9   DATA LOAD OPTIMIZATION STRATEGIES

- Tag resources for cost attribution, enabling granular visibility into where costs are incurred and who is responsible.

- Set budgets and alerts to prevent unexpected cost overruns.

- Review usage reports regularly for proactive optimization and accountability.

With these optimization principles in mind, it is time to explore their application in practice. The next section focuses on batch loading, one of the most scalable methods for structured data ingestion in Snowflake.

Batch Loading: Efficient Data Ingestion at Scale

In modern data platforms, ingesting large volumes of data efficiently and reliably is a foundational requirement. Snowflake supports multiple ingestion methods, but batch loading remains one of the most common and effective approaches for structured, periodic data ingestion. Whether loading data from cloud storage or integrating with external systems, batch loading provides a scalable and cost-effective solution.

Use the following query to identify large files that may be costly to load (the following image shows an example).

```
SELECT
    FILE_NAME,
    (FILE_SIZE / 1024 / 1024) AS FILE_SIZE_MB,
    STATUS,
    LAST_LOAD_TIME
FROM SNOWFLAKE.ACCOUNT_USAGE.COPY_HISTORY
WHERE
    FILE_SIZE > 1 * 1024 * 1024   -- Files larger than 250MB
    AND LAST_LOAD_TIME >= DATEADD(DAY, -30, CURRENT_TIMESTAMP())
ORDER BY FILE_SIZE DESC;
```

FILE_NAME	FILE_SIZE_MB	STATUS	LAST_LOAD_TIME
TPCH_SF10/LINEITEM/LINEITEM.csv_2_1_3.csv.gz	265.606369018555	Loaded	2025-06-26 13:53:34.079
TPCH_SF10/LINEITEM/LINEITEM.csv_2_6_4.csv.gz	264.421066284180	Loaded	2025-06-26 13:53:34.079
TPCH_SF10/LINEITEM/LINEITEM.csv_0_3_4.csv.gz	264.420898437500	Loaded	2025-06-26 13:53:34.079
TPCH_SF10/LINEITEM/LINEITEM.csv_2_5_4.csv.gz	264.416778564453	Loaded	2025-06-26 13:53:34.079
TPCH_SF10/LINEITEM/LINEITEM.csv_1_6_4.csv.gz	264.365768432617	Loaded	2025-06-26 13:53:34.079

Considerations

- Use optimal file sizes (100–250 MB compressed) to improve parallelism and performance; avoid very small or very large files.
- Monitor and troubleshoot failed loads using Snowflake load history and error views; consider staging tables to validate data before merging into production.
- Prevent data duplication by ensuring idempotency through file naming conventions, metadata tracking, or deduplication logic.
- Secure cloud storage locations with proper access controls and encryption; use Snowflake external stages with scoped credentials for secure access.
- Automate batch loads using orchestration tools and set up alerts or dashboards to monitor load success, latency, and volume trends.

While batch loading suits periodic data, modern needs often require real-time pipelines. For this, Snowflake offers Snowpipe, a serverless, automated solution for low-latency, continuous data ingestion from cloud storage.

Continuous Data Ingestion with Snowpipe

In the era of real-time analytics and data-driven decision-making, the ability to ingest data continuously and efficiently is critical. Snowpipe provides a robust, serverless solution for continuous data loading, enabling near real-time ingestion of data from cloud storage (e.g., Amazon S3, Azure Blob Storage, Google Cloud Storage) into

Snowflake tables. Unlike traditional batch-loading methods, Snowpipe, illustrated in Figure 9-2, allows data to be loaded in micro-batches as soon as it becomes available in a stage, significantly reducing latency and operational overhead. It is designed to support:

> **Micro-batch loading:** Data is loaded as soon as it is available, reducing the delay between data arrival and availability for querying.
>
> **Serverless architecture:** Snowflake manages the compute resources required for data ingestion.
>
> **Support for semi-structured data:** JSON, Avro, Parquet, and other formats are supported.

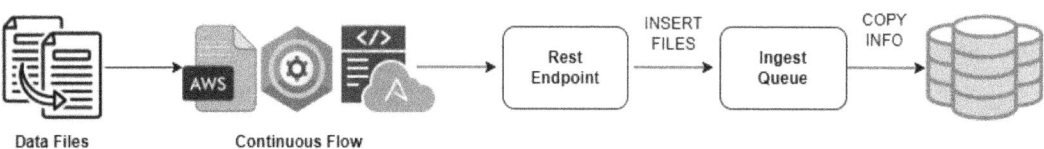

Figure 9-2. Snowpipe data load flow

At the core of Snowpipe is the pipe object, which encapsulates a [COPY INTO] SQL statement that defines how data should be loaded from a stage into a target table.

Automating Data Loads with Snowpipe

Snowpipe supports two primary methods for automating data ingestion: cloud messaging integration and manual REST API calls. The most efficient and scalable approach leverages cloud messaging, allowing Snowpipe to automatically detect and load new files as they arrive in cloud storage.

Cloud Messaging Integration

Snowpipe can be seamlessly integrated with event notification systems from major cloud providers. When a new file is uploaded to a designated stage, the cloud service sends an event notification to Snowpipe, which then triggers the data load process automatically. This eliminates the need for manual intervention and ensures near real-time data availability.

CHAPTER 9 DATA LOAD OPTIMIZATION STRATEGIES

Supported cloud providers and services:

- **Amazon S3:** Utilizes S3 event notifications in combination with AWS SNS (Simple Notification Service) and SQS (Simple Queue Service).

- **Azure Blob Storage:** Leverages Azure Event Grid to publish events when new blobs are created.

- **Google Cloud Storage:** Uses Google Pub/Sub to deliver notifications for new file uploads.

This event-driven architecture enables efficient, scalable, and low-latency data ingestion pipelines, making it ideal for streaming or frequently updated datasets.

REST API Integration

In scenarios where cloud event notifications are not feasible or when greater control over the ingestion process is required, Snowpipe offers a REST API integration method. This approach allows users to programmatically trigger data loads, making it suitable for custom workflows, batch processing, or environments with strict orchestration requirements.

The REST API method involves three key steps:

1. **Create a stage and pipe:** Define a named stage to store incoming files and create a Snowpipe object that maps to the target table and file format.

2. **Upload files to the stage:** Transfer data files to the designated stage using supported tools or SDKs.

3. **Trigger the load via REST API:** Call the Snowpipe REST endpoint, passing the names of the uploaded files. Snowpipe then processes the files and loads the data into the target table.

This method provides flexibility for integrating Snowflake into existing data pipelines, especially when precise control over timing, file selection, or error handling is required.

SQL Example: Creating a Pipe for Snowpipe

Step 1: Create a named stage (if not already created):

```
CREATE OR REPLACE STAGE web_stage
  URL='s3://weblog/data/'
  STORAGE_INTEGRATION = web_s3_integration;
```

Step 2: Create a pipe to automate data loading:

```
CREATE PIPE DEMO_SNOWPIPE
AUTO_INGEST = true
INTEGRATION = 'DEMO_INTEGRATION'
AS
COPY INTO DEMO_DB.WEBLOG
FROM @web_stage/weblog/
file_format = (type = csv field_delimiter = '|' skip_header = 1);
```

Once the pipe is created, Snowpipe will monitor the stage and load new files into the WEBLOG table automatically, provided that cloud messaging is configured or the REST API is used to trigger the load.

Key Benefits of Using Snowpipe

Snowpipe offers a powerful, automated solution for continuous data ingestion, making it easier to keep your analytics up to date with minimal effort.

- **Reduced latency:** Data becomes available for querying within minutes of being staged, enabling near real-time analytics.

- **Operational simplicity:** Snowpipe is fully managed and serverless—no need to provision or manage compute resources or schedule batch jobs.

- **Scalability:** Automatically scales with data volume, handling spikes in load without manual intervention.

- **Cost efficiency:** You pay only for the compute used during data loading, with no idle resource costs.

Snowpipe Cost Model

Snowpipe uses a serverless compute model, so you don't manage compute resources directly. Instead, Snowflake handles the compute and charges based on usage.

- **Compute cost:** Billed at 1.25x the rate of standard virtual warehouses. For example, using the equivalent of an X-Small warehouse costs 1.25 credits/hour.

- **File overhead fee:** 0.06 credits per 1,000 files loaded. This can add up quickly if files are too small or too numerous.

Snowpipe enables seamless, automated data ingestion with minimal configuration and maintenance. Its serverless architecture and support for both structured and semi-structured data make it an ideal solution for modern, scalable data pipelines. For real-time, high-frequency ingestion, Snowpipe offers unmatched convenience and efficiency. However, for bulk or infrequent data loads, using `COPY INTO` with virtual warehouses may provide better cost control and flexibility.

While Snowpipe enables continuous ingestion, downstream transformations often need orchestration. Using the Snowflake Dynamic Tables feature simplifies this by offering declarative, auto-refreshing pipelines. The next section explores how Dynamic Tables keeps data consistently up to date.

Dynamic Tables for Scalable Auto-Refreshing Pipelines

As data ecosystems grow more complex, the need for simplified, scalable, and automated data transformation becomes critical. Dynamic tables offer a modern solution to this challenge by enabling declarative, incremental data pipelines that automatically stay up to date. Unlike traditional ETL processes that require orchestration and scheduling, these tables are self-updating views that materialize data based on defined logic and dependencies.

This section explores the concept of dynamic tables, their benefits, practical use cases, and important considerations for implementation.

What Is a Snowflake Dynamic Table?

A *dynamic table* in Snowflake is a declarative object that materializes the result of a SQL query and automatically refreshes as the underlying data changes, as illustrated in Figure 9-3. Unlike traditional tables, dynamic tables do not require manual orchestration or complex merge logic.

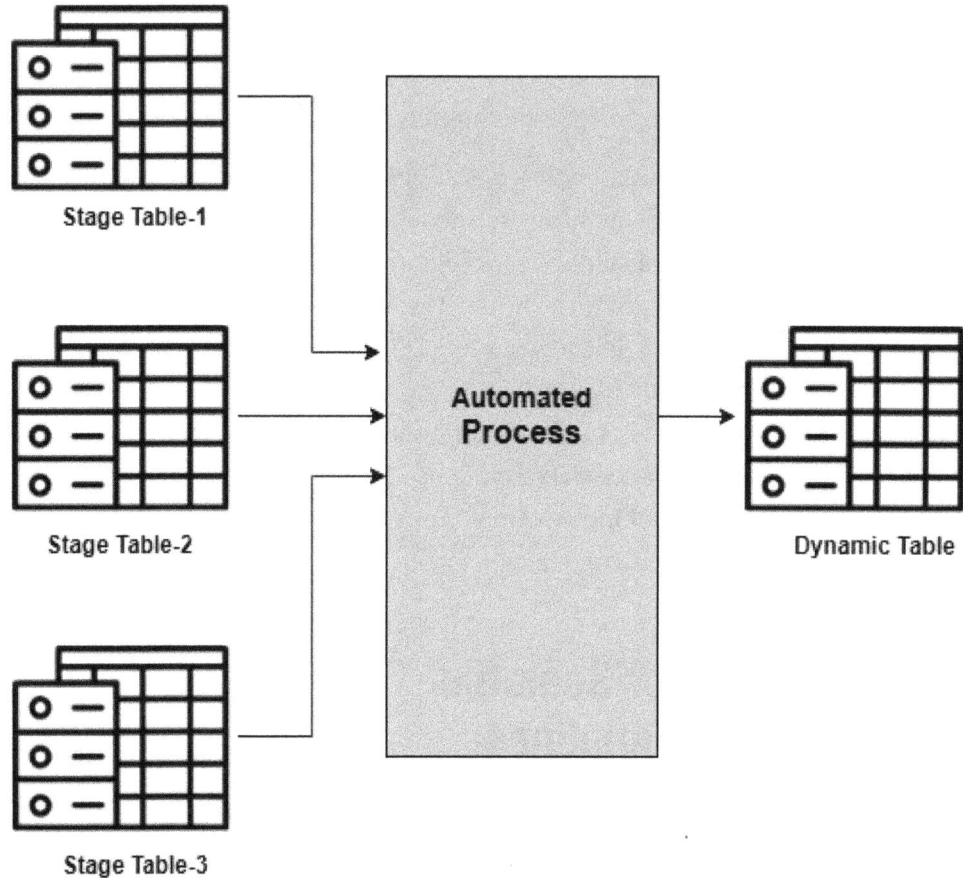

Figure 9-3. Dynamic table automated refresh process

Key Features

Dynamic tables in Snowflake offer built-in capabilities that streamline pipeline management while ensuring up-to-date data through automated, scalable transformations.

- Automatically track and apply changes from source tables.
- Support both FULL and INCREMENTAL refresh modes.
- Refresh frequency is controlled via the TARGET_LAG parameter.
- Execute using a specified virtual warehouse, impacting cost and performance.

Why Use Dynamic Tables?

Dynamic tables offer several advantages that simplify data pipeline development and enhance operational efficiency:

- Dynamic tables eliminate the need for streams, tasks, and complex scheduling by allowing a single SQL statement to define transformation logic, refresh cadence, and compute resource.
- Snowflake handles orchestration and refresh logic automatically, reducing the need for manual monitoring and error handling.
- With INCREMENTAL refresh mode, compute is only used when changes occur, minimizing unnecessary warehouse usage and aligning with FinOps principles.
- Dynamic tables support both batch and streaming use cases using the same SQL syntax, unifying pipeline development and reducing tooling overhead.

Dynamic Table Example

The following example demonstrates how to define a dynamic table with a daily refresh schedule and a full refresh mode:

```
CREATE DYNAMIC TABLE dt_sales_aggregate
  TARGET_LAG = '1 day'
  WAREHOUSE = DEMO_WH
  REFRESH_MODE = FULL
AS
```

```sql
SELECT s.sale_date, st.region, SUM(s.amount)
FROM sales s
JOIN stores st ON s.store_id = st.store_id;
```

Optimization Tips for FinOps

To maximize the cost-efficiency of dynamic tables, consider the following FinOps-aligned optimization practices:

- **Choose the right refresh mode:** Use INCREMENTAL for large datasets with small changes to reduce compute costs.

- **Set the appropriate** TARGET_LAG**:** Align the refresh frequency with business needs. Avoid overly aggressive lags that increase warehouse usage unnecessarily.

- **Monitor credit usage:** Use Snowsight or third-party tools like DataDog to track refresh frequency and warehouse consumption.

- **Optimize SQL logic:** Since the query runs on every refresh, ensure it is efficient, uses filters and partitions, and avoids unnecessary joins.

Dynamic tables offer a powerful, simplified, and cost-effective way to build and manage ELT pipelines in Snowflake. For FinOps practitioners, they provide a clear path to reducing operational overhead, improving data freshness, and optimizing compute spend. However, careful configuration of refresh modes, warehouse sizes, and monitoring practices is essential to fully realize their benefits.

While dynamic tables simplify transformations, challenges like data duplication and cross-platform access remain. Snowflake addresses this with Apache Iceberg, a table format that enables efficient, shared access across engines. The next section explores how Iceberg streamlines modern data architectures.

Apache Iceberg Tables: Eliminating Data Redundancies

As data volumes continue to grow and data architectures become increasingly complex, reducing redundancy in ETL pipelines is critical—not only for improving performance but also for enhancing cost efficiency. One effective strategy for achieving this is the adoption of Apache Iceberg tables, particularly when used with Snowflake support for externally managed Iceberg tables.

This section explores how organizations can streamline data ingestion, minimize storage duplication, and simplify transformation workflows by leveraging Iceberg as a centralized data layer. As illustrated in Figure 9-4, Iceberg tables are built on the Apache Iceberg open table format specification, which abstracts data files stored in open formats and provides robust support for:

- ACID transactions (atomicity, consistency, isolation, durability)
- Schema evolution to accommodate structural changes over time
- Hidden partitioning for simplified query optimization
- Table snapshots to enable Time Travel and rollback capabilities

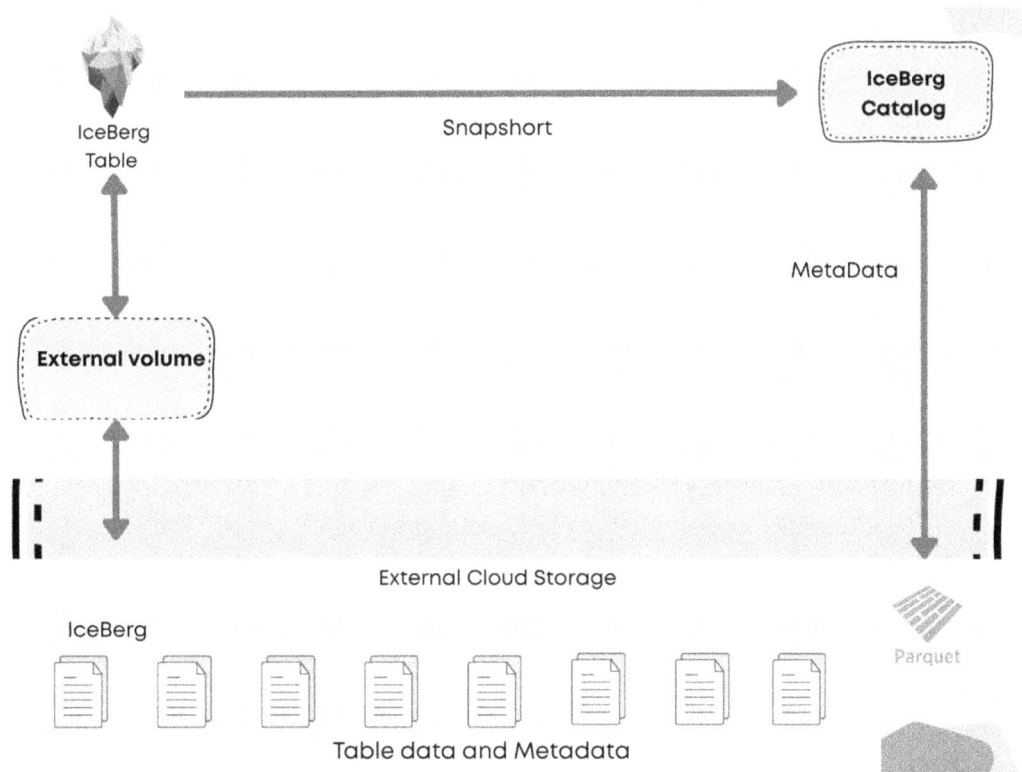

Figure 9-4. Snowflake Iceberg tables

The Problem: Redundant Data Movement

Traditional ETL pipelines often involve unnecessary duplication—data is extracted from a source, exported to cloud storage, and then reloaded into a data warehouse for transformation. This leads to:

- Higher storage costs from maintaining multiple copies
- Increased compute usage due to repeated ingestion
- Greater complexity in managing and maintaining pipelines

This inefficiency not only drives up operational costs but also slows down data delivery and increases the risk of inconsistencies.

The Solution: Iceberg as a Central Data Layer

Apache Iceberg offers a modern approach to simplifying data pipelines by acting as a central, authoritative data layer. Instead of duplicating data across multiple systems and formats, Iceberg allows data to be written once and accessed directly by multiple engines. This eliminates the need for intermediate exports and repeated ingestion steps.

Iceberg open table format supports:

- **Multi-engine access:** Query the same dataset from different platforms without moving data.
- **Schema evolution:** Adapt to changes in data structure without breaking pipelines.
- **Efficient metadata management:** Enable fast queries through partition pruning and lightweight metadata scans.

By adopting Iceberg, organizations can reduce redundancy, simplify architecture, and maintain a single source of truth—making data pipelines more efficient, scalable, and cost-effective.

Building a Modern Iceberg-Based Data Pipeline

A modern data pipeline built on Apache Iceberg is designed to reduce redundancy, simplify transformations, and support multi-engine analytics. By integrating cloud-native storage, external table definitions, and transformation tools, this architecture creates a scalable and efficient data layer. At the heart of this design are several key components:

- **Cloud object storage:** Serves as the central repository for Iceberg table data and metadata, enabling scalable, low-cost storage.

- **Data engines (e.g., data warehouses, query engines):** Access Iceberg tables directly from object storage without re-ingestion, supporting real-time and batch analytics.

- **External table definitions:** Allow engines to reference Iceberg tables using metadata catalogs, eliminating the need for data duplication.

- **Transformation tools (e.g., dbt):** Treat Iceberg tables as native sources, enabling seamless modeling and transformation workflows.

Maximizing Efficiency with Iceberg

Adopting Apache Iceberg as a central data layer offers a strategic advantage for organizations aiming to modernize their data pipelines. By eliminating redundant data movement and enabling direct access across multiple engines, Iceberg simplifies architecture while improving performance and cost efficiency. These advantages become even more apparent when we look at the core benefits it brings to modern data operations:

- **Cost efficiency:** Reduces storage and ingestion costs by eliminating duplicate data copies and unnecessary ETL steps.

- **Simplicity:** Streamlines pipelines by removing intermediate formats and reducing orchestration overhead.

- **Flexibility:** Supports multi-engine analytics (e.g., data warehouses, query engines, transformation tools) without additional data movement.

- **Scalability:** Leverages columnar storage and partition pruning to maintain high performance as data volumes grow.

CHAPTER 9　DATA LOAD OPTIMIZATION STRATEGIES

Best Practices for Implementation

To ensure a successful and efficient deployment of Iceberg tables, it is important to follow a set of strategic best practices that enhance performance, reduce costs, and maintain consistency across systems. Consider the following:

- **Co-locate data and compute:** Store Iceberg tables in the same region as your query engine to reduce latency and egress costs.

- **Automate metadata synchronization:** Use scheduled tasks or procedures to keep external engines in sync with the latest Iceberg metadata.

- **Optimize partitioning:** Design partition strategies that align with common query patterns to improve scan efficiency and reduce compute usage.

Adopting Apache Iceberg as a central data layer enables organizations to modernize their data pipelines, reduce operational costs, and build scalable, flexible architectures. By eliminating redundant data movement and embracing open standards, teams can achieve greater agility and align with FinOps goals for sustainable data operations.

Optimizing Snowflake virtual warehouses is key to boosting performance and controlling costs. The next section explores how to align configurations with workload and FinOps goals.

Tailoring Virtual Warehouse Size for Efficient Data Loading

Optimizing the size of Snowflake virtual warehouses (VWs) during data loading is a key FinOps strategy. While Snowflake elasticity allows for rapid scaling, choosing the right warehouse size ensures a balance between performance, cost-efficiency, and SLA adherence.

CHAPTER 9   DATA LOAD OPTIMIZATION STRATEGIES

Why It Matters

Snowflake COPY operations are capable of handling large-scale data ingestion, but their performance and cost-efficiency are closely tied to the size of the virtual warehouse. Over-provisioning can lead to unnecessary expenditure, while under-provisioning may result in slower load times and delays in critical data pipelines, as illustrated in Figure 9-5.

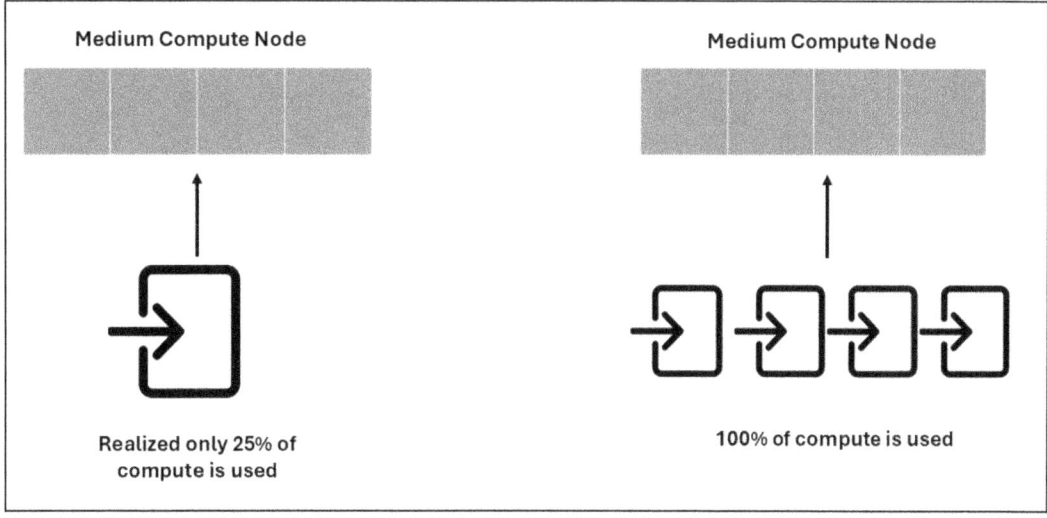

Figure 9-5. Data load parallel efficiency

These are the key metadata tables for analysis:

- COPY_HISTORY: Captures metadata about files loaded into Snowflake, including file size, row count, and load time. This helps in understanding the volume and structure of data ingested.

- QUERY_HISTORY: Logs execution details of COPY commands, such as execution time, bytes written, and warehouse size used. This is essential for correlating performance with resource usage.

CHAPTER 9 DATA LOAD OPTIMIZATION STRATEGIES

Methodology for Warehouse Sizing

Set thresholds and optimal parameters for choosing right virtual warehouses, as listed in Table 9-1.

Table 9-1. Warehouse Threshold

VWH Size	DB Size	Threads	File Count	Rec. Inbound Data Size (GB)	Rec. File Size (MB)
X-Small	1	8	24	4	250
Small	2	16	64	16	250
Medium	4	32	128	32	500
Large	8	64	256	64	750
X-Large	16	128	512	128	750
2X-Large	32	256	1024	256	1500
3X-Large	64	512	2048	512	1500

Database size: Refers to the number of clusters utilized to construct the virtual warehouse in Snowflake.

Threads: Snowflake designates eight threads or sessions per cluster. While COPY statements are generally not memory-intensive, it's crucial to consider this factor for optimal load performance. In this context, adhering to Snowflake's recommendation of using eight threads is advisable.

File count: This calculation involves multiplying the number of available threads by three for a small warehouse and by four for a larger warehouse. However, a high count of files has the potential to impact the elapsed time for critical SLA-driven jobs.

Recommended inbound data size: Signifies the advised data volume for loading by the virtual warehouse.

Recommended file size: Snowflake recommends inbound file sizes between 100 MB and 250 MB for optimal load throughput. However, accommodating this recommendation might be challenging for diverse data source systems. Consequently, many enterprises need to fine-tune their load strategy to balance throughput, time, and cost considerations.

For this analysis, an average file size ranging from 250 MB to 1500 MB has been chosen, considering the practical challenges associated with aligning with Snowflake's recommended file sizes.

Tracing Data Lineage: Utilizing Mental Math to Connect COPY_HISTORY and QUERY_HISTORY

The current metadata table lacks the necessary lineage details, creating an information gap. This deficiency poses challenges in comprehending the lineage between COPY_HISTORY and QUERY_HISTORY.

1. COPY_HISTORY furnishes information about filenames, sizes, and row counts loaded into specific tables. However, when multiple load jobs commence simultaneously, there lacks a clear classification. To address this, we aggregate data at the schema and table levels based on load time, discarding fractional seconds.

```
CREATE OR REPLACE TRANSIENT TABLE SF_COPY_HST
(
    SELECT  TABLE_SCHEMA_NAME,
            TABLE_NAME,
DATE_TRUNC('SECOND', LAST_LOAD_TIME) AS    LAST_LOAD_TIME_HR,
            SUM(CASE WHEN FILE_SIZE > 0 THEN FILE_SIZE/POWER(1024, 2)
                ELSE FILE_SIZE
                END) AS MB_FILE_SIZE_TOTAL,
            AVG(CASE WHEN FILE_SIZE > 0 THEN FILE_SIZE/POWER(1024, 2)
                ELSE FILE_SIZE
                END) AS AVG_MB_FILE_SIZE_TOTAL,
            MIN(CASE WHEN FILE_SIZE > 0 THEN FILE_SIZE/POWER(1024, 2)
                ELSE FILE_SIZE
                END) AS MIN_MB_FILE_SIZE_TOTAL,
            MAX(CASE WHEN FILE_SIZE > 0 THEN FILE_SIZE/POWER(1024, 2)
                            ELSE FILE_SIZE
                            END) AS MAX_MB_FILE_SIZE_TOTAL,
            SUM(ROW_COUNT) AS ROW_COUNT,
```

```
            SUM(ROW_PARSED) AS ROW_PARSED,
            COUNT(FILE_NAME) AS FILE_COUNT
    FROM SNOWFLAKE.ACCOUNT_USAGE.COPY_HISTORY
    WHERE LAST_LOAD_TIME::DATE >= CURRENT_DATE - 15
    GROUP BY ALL
);
```

2. The Query History captures details on COPY statement execution time, bytes loaded, and rows processed. Load jobs are classified using the query type, but notably, the table name is not independently logged in the query history. As a result, we extract the table name from the query text for comprehensive tracking.

```
CREATE OR REPLACE TRANSIENT TABLE SF_QRY_HST
(
SELECT  USER_NAME
        ,DATE_TRUNC('SECOND', START_TIME) AS START_TIME_HR
        ,START_TIME::DATE as THE_DT
        ,QUERY_ID
        ,UPPER(TRIM(SUBSTR((SUBSTR(QUERY_TEXT,POSITION('.', QUERY_TEXT,
        1)+1)),1,POSITION(' ', TABLE_STR, 1)))) AS TABLE_NAME
        ,QUERY_TYPE
        ,WAREHOUSE_SIZE
        ,WAREHOUSE_NAME
        ,EXECUTION_STATUS
        ,EXECUTION_TIME
        ,((EXECUTION_TIME/(60000))) AS EXEC_TIME_MIN
        ,ROWS_PRODUCED
        ,ERROR_MESSAGE
        ,CASE WHEN BYTES_WRITTEN > 0 THEN BYTES_WRITTEN/POWER(1024, 2)
            ELSE BYTES_WRITTEN
            END AS MB_WRITTEN
FROM SNOWFLAKE.ACCOUNT_USAGE.QUERY_HISTORY QH
    WHERE START_TIME::DATE >= CURRENT_DATE - 15
    AND USER_NAME='DATA_INGEST_USER'
```

```
        AND QUERY_TYPE='COPY'
        AND EXECUTION_TIME > 0
        AND WAREHOUSE_NAME IS NOT NULL
        AND CLUSTER_NUMBER IS NOT NULL
);
```

3. Create a lineage by establishing a connection between COPY_HISTORY and QUERY_HISTORY using load timestamps (discard fractional seconds) and table names extracted from query text. Identify the recommended warehouse based on the threshold and optimal parameters previously defined.

    Comparing the maximum of bytes written to Snowflake against bytes inbound from physical files is crucial. This is significant because often, the bytes written to the database are considerably higher than the inbound bytes, influenced by factors such as data skew and the impact of a large number of columns on the table header. Conversely, if the bytes written are less than the inbound data, this could be attributed to effective compression.

```
CREATE OR REPLACE TRANSIENT TABLE SF_VWH_ANLY
(
    SELECT a.USER_NAME             AS USER_NAME,
        a.START_TIME_HR            AS START_TIME,
        a.TABLE_NAME               AS TABLE_NAME,
        a.WAREHOUSE_NAME           AS WAREHOUSE_NAME,
        a.WAREHOUSE_SIZE           AS VWH_IN_USED,
        CASE
            WHEN a.WAREHOUSE_SIZE='X-Small' THEN 1
            WHEN a.WAREHOUSE_SIZE='Small' THEN 2
            WHEN a.WAREHOUSE_SIZE='Medium' THEN 3
            WHEN a.WAREHOUSE_SIZE='Large' THEN 4
            WHEN a.WAREHOUSE_SIZE='X-Large' THEN 5
            WHEN a.WAREHOUSE_SIZE='2X-Large' THEN 6
            WHEN a.WAREHOUSE_SIZE='3X-Large' THEN 7
            ELSE 0
        END AS VWH_IN_USE_NUM,
```

CHAPTER 9   DATA LOAD OPTIMIZATION STRATEGIES

```
    CASE
        WHEN (b.FILE_COUNT <= 24 AND GREATEST(a.MB_WRITTEN,b.MB_FILE_
        SIZE_TOTAL) <= 4*1024 AND b.AVG_MB_FILE_SIZE_TOTAL <= 250
        ) THEN 1
        WHEN ((b.FILE_COUNT <= 64 AND GREATEST(a.MB_WRITTEN,b.MB_FILE_
        SIZE_TOTAL) <= 16*1024) AND b.AVG_MB_FILE_SIZE_TOTAL <= 250
        ) THEN 2
        WHEN ((b.FILE_COUNT <= 128 AND GREATEST(a.MB_WRITTEN,b.MB_FILE_
        SIZE_TOTAL) <= 32*1024) AND b.AVG_MB_FILE_SIZE_TOTAL <= 500
        ) THEN 3
        WHEN ((b.FILE_COUNT <= 256 AND GREATEST(a.MB_WRITTEN,b.MB_FILE_
        SIZE_TOTAL) <= 64*1024) AND b.AVG_MB_FILE_SIZE_TOTAL <= 750
        ) THEN 4
        WHEN ((b.FILE_COUNT <= 512 AND GREATEST(a.MB_WRITTEN,b.MB_FILE_
        SIZE_TOTAL) <= 128*1024) AND b.AVG_MB_FILE_SIZE_TOTAL <= 750
        ) THEN 5
        WHEN ((b.FILE_COUNT <= 1024 AND GREATEST(a.MB_WRITTEN,b.MB_
        FILE_SIZE_TOTAL) <= 256*1024) AND b.AVG_MB_FILE_SIZE_TOTAL <=
        1500 ) THEN 6
        WHEN ((GREATEST(a.MB_WRITTEN,b.MB_FILE_SIZE_TOTAL) > 256*1024)
        AND b.AVG_MB_FILE_SIZE_TOTAL <= 750 ) THEN 7
      ELSE
         11
    END                                    AS VWH_RECOMMENDED,
    a.MB_WRITTEN::DECIMAL(12,2)            AS MB_WRITTEN,
    a.ROWS_PRODUCED::INTEGER               AS ROWS_WRITTEN,
    b.MB_FILE_SIZE_TOTAL::DECIMAL(12,2)    AS MB_FILE_SIZE_TOTAL,
    b.ROW_COUNT::INTEGER                   AS ROWS_FR_FILE,
    b.FILE_COUNT::INTEGER                  AS FILE_COUNT,
    a.EXEC_TIME_MIN::DECIMAL(12,2)         AS EXEC_TIME_MIN,
    a.EXECUTION_STATUS                     AS EXECUTION_STATUS,
    b.MIN_MB_FILE_SIZE_TOTAL::DECIMAL(32,2)   AS MIN_MB_FILE_
    SIZE_TOTAL,
    b.MAX_MB_FILE_SIZE_TOTAL::DECIMAL(32,2)   AS MAX_MB_FILE_
    SIZE_TOTAL,
```

CHAPTER 9   DATA LOAD OPTIMIZATION STRATEGIES

```
      b.AVG_MB_FILE_SIZE_TOTAL::DECIMAL(32,2)    AS AVG_MB_FILE_SIZE_TOTAL,
a.QUERY_ID,
      CASE WHEN  b.TABLE_NAME IS NULL THEN 'COPY_SESSION_NO_MATCH' ELSE
'COPY_SESSION_MATCH' END AS COPY_SESSION
FROM SF_QRY_HST a LEFT OUTER JOIN SF_COPY_HST b
ON a.TABLE_NAME = b.TABLE_NAME
AND a.START_TIME_HR = b.LAST_LOAD_TIME_HR
WHERE b.TABLE_NAME IS NOT NULL
);
```

4. Generate the summary of the optimal virtual warehouse for each table. When the same table is loaded using multiple virtual warehouses, distinct records are tracked to distinguish the elapsed time.

```
CREATE OR REPLACE TRANSIENT TABLE SF_VWH_REC
(
SELECT USER_NAME
      ,TABLE_NAME
      ,VWH_IN_USE_NUM              AS IN_USE_VWH
      ,MIN(VWH_RECOMMENDED)        AS MIN_VWH_SZ
      ,MAX(VWH_RECOMMENDED)        AS MAX_VWH_SZ
      ,ROUND(AVE(VWH_RECOMMENDED)) AS REC_VWH_SZ
      ,COUNT(QUERY_ID)             AS JOB_COUNT
      ,MIN(START_TIME::DATE)       AS MIN_DATE
      ,MAX(START_TIME::DATE)       AS MAX_DATE
      ,AVG(MB_WRITTEN)             AS AVG_MB_WRITTEN
      ,AVG(ROWS_WRITTEN)           AS AVG_ROWS_WRITTEN
      ,AVG(ROWS_FR_FILE)           AS AVG_ROWS_FR_FILE
      ,AVG(MB_FILE_SIZE_TOTAL)     AS AVG_MB_FILE_SIZE_TOTAL
      ,AVG(FILE_COUNT)             AS AVG_FILE_COUNT
      ,AVG(EXEC_TIME_MIN)          AS AVG_EXEC_TIME_MIN
      ,MAX(AVG_MB_FILE_SIZE_TOTAL) AS MAX_MB_FILE_SIZE_TOTAL
   FROM SF_VWH_ANLY
   GROUP BY ALL
);
```

CHAPTER 9 DATA LOAD OPTIMIZATION STRATEGIES

    5. Establish an interim table for mapping virtual warehouses.

```
CREATE OR REPLACE TRANSIENT TABLE SF_VWH_SIZE
(
  VWH_NO     NUMBER,
  VWH_SIZE   VARCHAR NOT NULL,
  VWH_WEIGHT NUMBER NOT NULL
);

INSERT INTO SF_VWH_SIZE (VWH_NO, VWH_SIZE, VWH_WEIGHT)
VALUES  (1,'X-Small',1),
        (2,'Small',2),
        (3,'Medium',4),
        (4,'Large',8),
        (5,'X-Large',16),
        (6,'2X-Large',32),
        (7,'3X-Large',64),
        (8,'4X-Large',128),
        (9,'5X-Large',256),
        (10,'6X-Large',512)
        (11,'Req.Analysis',0)
;
```

    6. Generate a conclusive report featuring recommended virtual warehouse details and other pertinent metrics for informed decision-making. Load jobs exceeding the 3X-Large warehouse capacity have been categorized for "required analysis," indicating a need for closer examination due to potential factors such as exceptionally large file sizes or significantly voluminous datasets.

```
WITH CTE_TBL_CNT AS
(
   SELECT TABLE_NAME,COUNT(*) AS TABLE_CNT
   FROM SF_VWH_REC
   GROUP BY 1
```

```
)
SELECT  USER_NAME,
        r.TABLE_NAME,
        t.TABLE_CNT,
        MIN_DATE,
        MAX_DATE,
        w.VWH_SIZE                           AS VWH_IN_USE,
        v.VWH_SIZE                           AS VWH_RECOM,
        JOB_COUNT                            AS JOB_COUNT,
        AVG_MB_WRITTEN::DECIMAL(12,2)        AS MB_WRITTEN_TOTAL,
        AVG_MB_FILE_SIZE_TOTAL::DECIMAL(12,2) AS MB_FILE_SIZE_TOTAL,
        MAX_MB_FILE_SIZE_TOTAL::DECIMAL(12,2) AS MB_AVG_FILE_SIZE,
        AVG_ROWS_WRITTEN::NUMBER             AS AVG_ROWS_WRITTER,
        AVG_ROWS_FR_FILE::NUMBER             AS AVG_ROWS_FR_FILE,
        AVG_FILE_COUNT::NUMBER               AS AVG_FILE_COUNT,
        AVG_EXEC_TIME_MIN::DECIMAL(12,2)     AS AVG_TIME_MIN
    FROM SF_VWH_REC r INNER JOIN CTE_TBL_CNT t
        ON r.TABLE_NAME = t.TABLE_NAME
    LEFT OUTER JOIN SF_VWH_SIZE w
        ON r.IN_USE_VWH = w.VWH_NO
    LEFT OUTER JOIN SF_VWH_SIZE v
        ON r.REC_VWH_SZ = v.VWH_NO
GROUP BY ALL
;
```

7. The virtual warehouse recommendation result is shown in Table 9-2. While these results may not reflect the optimal outcomes for all scenarios, we recommend adjusting the threshold numbers based on your specific environment.

Table 9-2. Warehouse Recommendation

TABLE_NME	WH_IN_USE	WH_RECOM	JOB_COUNT	FILE_SIZE_MB
PRODUCT	Medium	X-Small	14	3,762
STORE	Medium	X-Small	1	0.01
RETURNS	Large	X-Small	15	306
DELIVERY	Medium	Medium	12	21,742
PAYMENT	Medium	Medium	12	4,976
PROMOTION	Medium	Large	13	62,713
ORDERS	Medium	X-Large	15	9,567

Summary

Snowflake offers a robust platform for scalable and cost-effective data ingestion, supporting diverse use cases, from analytics to machine learning. Success depends not just on speed, but on delivering fresh, accurate data aligned with business goals. By optimizing file sizes, tuning virtual warehouses, leveraging parallelism, and monitoring performance, teams can enhance efficiency and control costs. A well-optimized ingestion strategy ensures a responsive, future-ready data infrastructure.

With a comprehensive understanding of Snowflake data load optimization techniques, from batch and continuous ingestion to architectural enhancements like Iceberg and dynamic tables, teams are well-equipped to build efficient, scalable pipelines. As we shift focus to the next chapter, we explore how to optimize the Snowflake cloud services layer, including resource management, service scaling, and cost governance, to further enhance platform performance and operational efficiency.

CHAPTER 10

Cloud Services Optimization Strategies

Overview

This chapter focuses on optimizing the cloud services layer, also known as Snowflake's Global Service Layer (GSL). Acting as the brain of the platform, GSL orchestrates query engine activities, from receiving a query to navigating the optimizer, accessing storage, and returning results. It handles user authentication, query compilation, and performance enhancements like result caching. Additionally, it ensures smooth code rollouts and rollbacks, maintaining system stability and responsiveness.

To ensure these critical operations run efficiently and reliably, the cloud services layer incorporates a range of built-in mechanisms designed to optimize performance and resource utilization. These include request throttling and intelligent load-balancing across different availability zones, which help maintain consistent performance under varying workloads. Seamless data placement further enhances system responsiveness by efficiently distributing tasks across the cloud infrastructure.

Another foundational element in optimizing cloud services is the use of metadata, or the data dictionary. This centralized repository contains vital information about data access, usage, and relationships between objects. A well-maintained data dictionary not only supports better data governance and quality but also enables more efficient query compilation and execution, key factors in maintaining high performance and responsiveness.

Building on this understanding of how the cloud services layer functions, the following sections delve into specific optimization strategies that can be applied to further enhance operational efficiency and cost-effectiveness.

CHAPTER 10   CLOUD SERVICES OPTIMIZATION STRATEGIES

The essential optimization strategies covered in this chapter include:

- Understanding the cloud service layer architecture and tracking credit and dollar usage

- Recognizing the patterns driving cloud services usage and recommendations to handle them

- Learning the metadata layer limitations and best practices for using catalog data

- Recognizing the causes of metadata operations throttling and how to reduce it

- Reducing data dictionary size and query complexity to minimize throttling times on the Snowflake data dictionary using a persistent metadata layer

To learn more about operational efficiency and cost management with the cloud service layer, continue to the next session.

Understanding the Cloud Service Layer Architecture

Understanding the components within the cloud service layer is crucial, as illustrated in Figure 10-1. This architecture is characterized by a self-managed, no-knobs system, which offers several key advantages. Each subsystem can evolve independently, ensuring flexibility and adaptability. The health of virtual machines is managed separately from the autoscaling process, enhancing reliability. Additionally, the architecture is capable of operating at the scale required by Snowflake, demonstrating its efficiency and scalability.

CHAPTER 10 CLOUD SERVICES OPTIMIZATION STRATEGIES

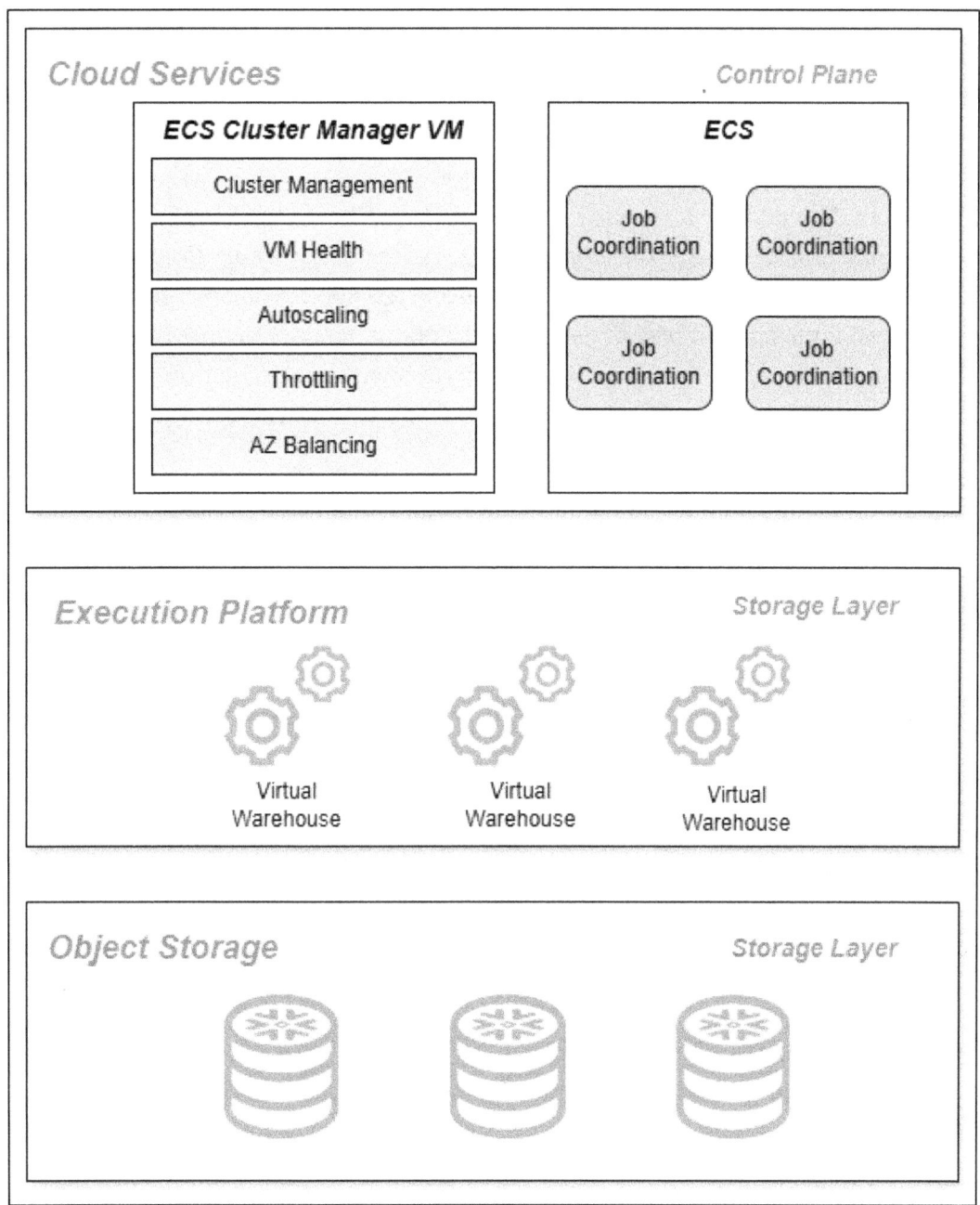

Figure 10-1. The Snowflake elastic cloud services architecture

Transitioning to the cloud service layer highlights the importance of managing cloud services and the associated costs. Use of cloud services is charged only if the daily consumption exceeds 10% of the daily use of virtual warehouses. This charge is calculated daily (in the UTC time zone) to ensure accurate application of the 10% adjustment at the credit price for that day.

If your daily cloud services usage is less than 10% of the compute credits used by your warehouses, Snowflake adjusts your charges so that you only pay for the actual cloud services used that day. Because this adjustment is applied daily, your total cloud services cost for the month may be less than 10% of your total warehouse usage. It is important to note that serverless features such as query acceleration, automatic clustering, materialized views, and search optimization are not included in this 10% adjustment.

Note Refer to Table 3-2 in Chapter 3 for an example of how cloud services credits are calculated.

To analyze and optimize cloud services usage, it is essential to monitor usage patterns and consider the associated recommendations if usage is higher than expected.

Monitoring Cloud Services Usage

To effectively manage your cloud services usage, you can query the catalog views to view detailed information at any point during the month.

In the WAREHOUSE_METERING_HISTORY view, query three columns to understand your total credit usage for compute and cloud services: Credits_Used, Credits_Used_Cloud_Services, and Credits_Used_Compute.

The total credits used are calculated as follows:

> Credits_Used = Credits_Used_Cloud_Services + Credits_Used_Compute.

This query will provide you with a breakdown of credit usage, including compute and cloud service-related credits, for the specified time period (the following image shows an example).

```sql
-- Cloud Service Credit Usage at Warehouse Level
SELECT
    WAREHOUSE_NAME,
    SUM(CREDITS_USED) AS CREDITS_USED,
    SUM(CREDITS_USED_CLOUD_SERVICES) AS CREDITS_USED_CLOUD_SERVICES,
    SUM(CREDITS_USED_COMPUTE) AS CREDITS_USED_COMPUTE
FROM
    SNOWFLAKE.ACCOUNT_USAGE.WAREHOUSE_METERING_HISTORY
WHERE CREDITS_USED_CLOUD_SERVICES > 0
    AND START_TIME >= DATE_TRUNC('MONTH', DATEADD(month,-3,CURRENT_
    TIMESTAMP()))
GROUP BY WAREHOUSE_NAME
ORDER BY CREDITS_USED_CLOUD_SERVICES DESC;
```

WAREHOUSE_NAME	CREDITS_USED	CREDITS_USED_CLOUD_SERVICES	CREDITS_USED_COMPUTE
TPCDS_BENCH_10T	21.077838949	0.019505565	21.058333384
TPCDS_BENCH_100T	13.645072513	0.015072502	13.630000011
COMPUTE_WH	0.914672495	0.000228056	0.914444439
CLOUD_SERVICES_ONLY	0.000125001	0.000125001	0.000000000

In the METERING_DAILY_HISTORY view, you can query the same usage information as in WAREHOUSE_METERING_HISTORY, with additional columns for Credits_Adjustment_Cloud_Services and Credits_Billed. This view enables you to see the daily adjustment received (which offsets your cloud services usage up to 10% of compute credits) and the net credits billed.

The calculation is as follows:

> Credits_Used + Credits_Adjustment_Cloud_Services = Credits_Billed

Note that the adjustment is a negative number.

This query will give you the daily breakdown of credit adjustments related to cloud services and the total credits billed during the specified period (the following image shows an example).

CHAPTER 10   CLOUD SERVICES OPTIMIZATION STRATEGIES

```
-- Breakdown of credit adjustments related to Cloud Services
SELECT
    usage_date,
    credits_used_cloud_services,
    credits_adjustment_cloud_services,
    credits_used_cloud_services + credits_adjustment_cloud_services AS
    billed_cloud_services
FROM SNOWFLAKE.ACCOUNT_USAGE.METERING_DAILY_HISTORY
WHERE usage_date >= DATE_TRUNC('MONTH', DATEADD(month,-3,CURRENT_
TIMESTAMP()))
    AND credits_used_cloud_services > 0
ORDER BY 4 DESC;
```

USAGE_DATE	CREDITS_USED_CLOUD_SERVICES	CREDITS_ADJUSTMENT_CLOUD_SERVICES	BILLED_CLOUD_SERVICES
2025-04-06	0.000108612	0.0000000000	0.0001086120
2025-04-14	0.034822512	-0.0348225120	0.0000000000

In the QUERY_HISTORY view, you can see a breakdown of job-level information, such as query ID, query text, start time, end time, and more. An additional column, called Credits_Used_Cloud_Services, shows the cloud services credits used by each individual query.

This query will give you details on the usage of cloud services credits for each query that was executed within the specified date range (the following image shows an example).

```
-- Cloud Services credits for each query
SELECT
    QUERY_ID,
    WAREHOUSE_NAME,
    USER_NAME,
    START_TIME,
    CREDITS_USED_CLOUD_SERVICES
FROM
    SNOWFLAKE.ACCOUNT_USAGE.QUERY_HISTORY
```

```
WHERE
    CREDITS_USED_CLOUD_SERVICES IS NOT NULL
    AND START_TIME >= DATE_TRUNC('MONTH', DATEADD(month,-3,CURRENT_
    TIMESTAMP()))
ORDER BY CREDITS_USED_CLOUD_SERVICES DESC;
```

QUERY_ID	WAREHOUSE_NAME	USER_NAME	START_TIME	CLOUD_SERVICES_CREDITS
01bbacdc-0002-af25-0007-b60e000212	TPCDS_BENCH_100T	VELNATSF	2025-04-13 20:40:16.393 -070(	0.009657
01bbb1f8-0002-b35e-0007-b60e0002e(	TPCDS_BENCH_10T	VELNATSF	2025-04-14 18:28:09.103 -070(	0.0039
01bbace1-0002-af4a-0007-b60e000223	TPCDS_BENCH_100T	VELNATSF	2025-04-13 20:45:17.838 -070(	0.003074
01bbb1f7-0002-b362-0007-b60e00030(	TPCDS_BENCH_10T	VELNATSF	2025-04-14 18:27:23.312 -070(	0.002927
01bbb1f5-0002-b35b-0007-b60e00029(	TPCDS_BENCH_10T	VELNATSF	2025-04-14 18:25:53.383 -070(	0.002925

It is important to note that the adjustment for included cloud services (10% of compute) is shown only on the capacity usage statement and in the METERING_DAILY_HISTORY view. The information viewable in the UI—WAREHOUSE_METERING_HISTORY view and QUERY_HISTORY view—does not account for this adjustment and will, therefore, be greater than your actual credit consumption.

Patterns Driving Cloud Services Usage

If you find that your cloud services usage is higher than expected, it is important to review the following patterns, which drive cloud services usage. These insights will help you assess whether adjustments are needed to manage your costs effectively.

Pattern 1: Copy Commands with Poor Selectivity

Running copy commands with poor selectivity can result in significant time spent listing files from the cloud storage layer (e.g., AWS S3). This process utilizes only the cloud services layer.

Recommendation: Consider restructuring your S3 bucket to include a date prefix, allowing you to list only the targeted files you need.

For example, instead of using a generic path like this:

```
COPY INTO my_table FROM 's3://sales-bucket/data/';
```

Use a more selective path with a date prefix to limit the number of files scanned, like so:

```
COPY INTO my_table FROM 's3://sales-bucket/data/2025/07/12/';
```

Pattern 2: High Frequency DDL Operations or Cloning

Data Definition Language (DDL) operations, particularly cloning, are entirely metadata operations, meaning that they consume only cloud services resources. Frequently creating or dropping large schemas or tables, or cloning databases for backup, can result in significant cloud services usage.

Recommendation: Cloning uses only a fraction of the resources needed for deep copies, so we encourage you to continue utilizing this capability. Review your cloning patterns to ensure they are at the granularity and frequency you intend. For example, consider cloning only the individual tables you need rather than the entire schema.

Pattern 3: High Frequency, Simple Queries

While any individual simple query may drive negligible cloud services consumption, running queries such as `SELECT 1`, `SELECT sequence.nextval`, or `SELECT current session` at an extremely high frequency (tens of thousands per day) can result in significant cloud services usage in aggregate.

Recommendation: Review your query frequency to ensure that it is appropriately set for your use case. If you observe a high frequency of `SELECT current session` queries originating from partner tools using the JDBC driver, confirm that the partner has updated their code to use the `getSessionId()` method in the `SnowflakeConnection` interface. This method leverages caching and reduces cloud services usage.

Pattern 4: High Frequency Information_Schema Queries

`Information_Schema` commands consume only cloud services resources. While any individual query may drive negligible cloud services consumption, running these queries at an extremely high frequency (tens of thousands per day) can result in significant cloud services usage in aggregate.

Recommendation: Review your query frequency to ensure it is appropriately set for your use case. Alternatively, query the corresponding view in the Account_Usage share. Querying the Account_Usage share will use virtual warehouse (compute) credits rather than cloud services.

Pattern 5: Result Scan

If your workload primarily consists of a stream of queries that can be satisfied by the result cache, you will experience performance and cost benefits due to Snowflake's use of the cloud services layer. This is particularly advantageous for complex queries that typically require significant time to compile and execute on the data warehouse.

Recommendation: No change is necessary. In this scenario, you are achieving better value by satisfying these queries through the cloud services layer rather than the virtual warehouse (compute) layer. Utilizing the result cache is significantly faster and more cost-effective than relying on the data warehouse for every query.

Pattern 6: High Frequency SHOW Commands

SHOW commands, often seen with data applications or third-party tools, are entirely metadata operations and consume only cloud services resources.

Recommendation: This pattern typically occurs when an application built on top of Snowflake uses a large number of SHOW commands. These commands may also be initiated by third-party tools. Review your query frequency to ensure it is appropriately set for your use case. If partner tools are involved, reach out to your partner to enquire about any plans to adjust their usage.

Pattern 7: Single Row Inserts and Fragmented Schemas

When using a regular (columnar) table to store data, single row inserts and fragmented schemas can be suboptimal and incur significant cloud service usage. If you are building a data application and define a schema per customer, your data load can drive heavy cloud services consumption due to the frequency of loads for each customer within a given time period. This approach also results in a substantial amount of metadata that Snowflake needs to maintain. While each metadata operation individually consumes minimal resources, the aggregate consumption can be significant.

Recommendation: To optimize your data load process, consider using a hybrid table. If hybrid tables are not supported in your region or do not meet your needs, it is generally more efficient to perform batch or bulk loads rather than single row inserts. Implementing a shared schema can create significant efficiencies and cost savings. Clustering all tables on customer_ID and using secure views is recommended. If you choose not to change this pattern, be aware that you may incur additional cloud services usage charges to support the schema-per-customer approach.

Pattern 8: Complex SQL Queries

Several types of queries are particularly taxing on the cloud services layer, including those with numerous joins or Cartesian products, very large SQL queries, and extensive in-lists. These queries are characterized by high compilation times.

Recommendation: Review your queries to ensure they are performing as intended. Snowflake supports these complex queries and will charge you only for the resources consumed.

Did you know that a data dictionary can sometimes struggle with processing many objects in a query? This can lead to slowdowns, delays, and even overheads. Imagine it like a traffic jam—too many cars (or database objects) trying to move at once can cause congestion and slow everything down.

In the next section, we explore these intricate scenarios and provide valuable tips and techniques to mitigate these issues. By understanding and addressing these challenges, you can keep your database running smoothly and efficiently.

Exploring Metadata Layers and Best Practices

When working with metadata in Snowflake, there are several options available for accessing and retrieving it. The key is to choose the right one based on your specific use case and requirements. Here's a breakdown of the options:

- INFORMATION_SCHEMA **for smaller catalogs with fresh data:** This is ideal when you need to query smaller catalogs with a high degree of freshness. It supports full SQL queries, allowing you to filter, sort, group, and join data just like any other table. This option is perfect for obtaining up-to-date information and performing complex queries with ease.

- **SHOW commands for medium-sized catalogs:** When working with medium-sized catalogs that don't require complex queries, SHOW commands are an excellent choice. They provide recently updated data and support simple filtering. If your use case doesn't involve heavy querying, SHOW commands offer an efficient way to access metadata.

- **ACCOUNT_USAGE for large catalogs:** When working with large catalogs where some degree of staleness is acceptable, ACCOUNT_USAGE is a useful option. Although it is not as real-time as other methods, it is ideal for querying vast amounts of metadata when you can tolerate some delay in the results. Just remember, you will need the appropriate privileges to access this data.

As shown in Table 10-1, by understanding these options, you can choose the catalog that best fits your needs. Whether you require recently updated data, the flexibility of complex queries, or the ability to handle large volumes of metadata, there is a solution for you. Consider your specific use case and the importance of freshness, complexity, and volume to make an informed decision.

Table 10-1. Snowflake Catalog Usage and Limitations

	INFORMATION_SCHEMA	SHOW	ACCOUNT_USAGE
Size limitation	Unlimited	10k	Unlimited
Query support	Complex queries	Simple queries	Complex queries
Result freshness	Accurate	Accurate	Stale (min to hours)
Performance	Slow, single-threaded	Medium	Fast, multi-threaded
Fine grained RBAC	Yes	Yes	No
Cloud services resource	Yes	Yes	No
Warehouse required?	No	No	Yes

These layers address several key areas to help manage operations and optimize your Snowflake experience:

Metadata exploration: This layer is crucial for efficient database management and development. It offers a clear view of the database's structure through views like TABLES, COLUMNS, and SCHEMAS, making it easier to understand how your data is organized.

Security management: Snowflake enhances security and compliance by providing views such as USERS and ROLES, allowing you to track user privileges and access controls. This helps ensure that only authorized users can access and modify sensitive data.

Performance monitoring: To troubleshoot performance issues and optimize resource usage, Snowflake offers views like QUERY_HISTORY and WAREHOUSE_UTILIZATION. These help identify bottlenecks and ensure that resources are being used efficiently.

Cost tracking: Snowflake provides detailed usage information about compute, storage, and serverless services. This allows you to track consumption, project future usage, and set budgeting limits for better cost control.

The next section explores the role of the data dictionary in managing metadata and explains how a poorly managed data dictionary can lead to increased GSL (Global System Load) overhead.

Understanding Metadata Operations Throttling

Metadata operations that throttle in Snowflake's architecture can substantially affect both performance and efficiency. To optimize system operations effectively, it is crucial to comprehend the primary factors contributing to throttling. Key causes include many queries that require frequent metadata requests, which can strain system resources. This also puts significant stress on the GSL.

Understanding these factors is essential to implementing strategies that mitigate throttling and enhance overall system performance.

- **Metadata operations:** Metadata operations in Snowflake, including the use of SHOW commands and INFORMATION_SCHEMA views, can lead to performance and throttling issues. These operations access Snowflake's metadata services and are independent of warehouse or customer data storage. However, poorly constructed queries can result in significant delays or errors, such as "Information schema query returned too much data. Please repeat the query with more selective predicates."

- **High-frequency queries:** Simple queries (e.g., SELECT 1, SELECT sequence1.NEXTVAL, SELECT CURRENT_SESSION()), queries against INFORMATION_SCHEMA views, and SHOW commands at high frequency can significantly increase cloud services usage. These operations are metadata-only and consume cloud services resources.

- **Long compilation time:** When a significant portion of the total execution time is spent on compilation, it indicates that the query is highly dependent on metadata. Additionally, complex queries, particularly those involving multiple joins, subqueries, or extensive data processing, can consume more resources and lead to throttling.

- **Size of the data dictionary:** The data dictionary contains metadata about the database objects. As the size of the data dictionary grows, it can lead to increased overhead and slower query compilation time.

To identify queries with long compilation times, use the following query. A high number of such queries indicates the need to perform housekeeping on the data dictionary and validate recurring queries using SHOW commands and INFORMATION_SCHEMA views.

```
-- Identify queries with long compilation times
SELECT
 compilation_time,
 execution_time,
 (compilation_time / execution_time) as compilation_ratio
```

```
FROM account_usage.query_history
WHERE (compilation_time / execution_time) > 0.5
AND START_TIME >= DATE_TRUNC('DAY', DATEADD(day,-7,CURRENT_TIMESTAMP()))
ORDER BY compilation_ratio DESC
LIMIT 100;
```

# COMPILATION_TIME	# EXECUTION_TIME	# COMPILATION_RATIO
6010	10790	0.556997
14618	6345	2.303861
4353	5067	0.859088
1729	3151	0.548715
6129	3117	1.966314

For example, the following query uses the query_hash value to identify the query IDs for the top 100 recurring queries (the following image shows an example).

```
-- Use query_hash to identify top 100 recurring queries
SELECT
    query_hash,
    warehouse_name,
    COUNT(*) as query_count,
    SUM(total_elapsed_time) as total_elapsed_time,
    MAX(query_id) as max_queryid
FROM SNOWFLAKE.ACCOUNT_USAGE.QUERY_HISTORY
WHERE START_TIME >= DATE_TRUNC('DAY', DATEADD(day,-7,CURRENT_TIMESTAMP()))
GROUP BY query_hash
ORDER BY SUM(total_elapsed_time) DESC
LIMIT 100;
```

A QUERY_HASH	A WAREHOUSE_NAME	# QUERY_COUNT	# TOTAL_ELAPSED_TIME
f8068e7ed457bdb539322faf9e03639	TPCDS_BENCH_10T	53	4711
9ddc24dcc7b36224eabdf604aa2472t	COMPUTE_WH	41	3437
e6f8c5cfa7cbcc66fc2cea9209f5f722	TPCDS_BENCH_10T	28	9776
b39af5871b6b871f0d8dec946d99c4:	TPCDS_BENCH_10T	27	1914
4f57fbb0d3bfeafdb6f1b2bde9d29b50	TPCDS_BENCH_10T	27	1665

To mitigate the risk of metadata throttling in Snowflake, it is essential to adopt effective strategies. The next section describes some best practices you should follow to achieve optimal metadata efficiency:

Reducing Metadata Operations Throttling

Reducing metadata operations throttling is crucial to ensure that queries do not spend excessive time in the compilation step, which could affect the overall metadata operation and increase costs on cloud service resources.

Consider these key areas for optimization:

- **Limit result sets and add filters:** Apply more filters in the WHERE clause for INFORMATION_SCHEMA objects to limit the result set.

- **Avoid LIKE/ILIKE operators:** Refrain from using LIKE/ILIKE operators in the WHERE clause or ON conditions in INFORMATION_SCHEMA queries.

- **Separate filters for multiple objects:** If you're accessing multiple INFORMATION_SCHEMA objects in a single query, add separate filters for each view.

- **Use SHOW commands with fully qualified names:** For optimal performance, use SHOW commands with fully qualified names, but be aware of the limitation of 10,000 records.

- **JDBC driver optimization:** For high-frequency SELECT CURRENT_SESSION() queries originating from partner tools using the JDBC driver, ensure that the partner employs the getSessionId() method within the SnowflakeConnection interface. This approach leverages caching mechanisms to minimize usage.

- **Use SNOWFLAKE.ACCOUNT_USAGE:** For larger result sets and complex queries, utilize SNOWFLAKE.ACCOUNT_USAGE views, which have latency from 30 minutes to three hours, depending on the view. These views are not dependent on metadata services and use the user warehouse for compute.

The next section delves into metadata usage for optimization and explores detailed implementation strategies, considering the benefits and factors to evaluate before proceeding.

Reducing Data Dictionary Size and Query Complexity

To mitigate extended throttling times for Snowflake data dictionary (DD) queries, it is advisable to reduce the size of the data dictionary by eliminating unnecessary objects and utilizing temporary or transient tables for short-term data instead of permanent tables. Additionally, you can simplify query complexity by minimizing joins in queries on INFORMATION_SCHEMA and applying highly selective filters. Given that throttling imposes increasingly severe penalties on larger queries, executing multiple smaller queries rather than fewer larger ones may decrease the likelihood of throttling.

Maintaining a database free from redundant objects not only aids in reducing storage costs, but also enhances governance.

However, reengineering queries to break them into smaller components may require additional programming effort. Adopting SQL best practices from the outset of a project can alleviate the need for such rework later on, resulting in more efficient query performance and resource management.

Use ACCOUNT_USAGE

Begin by identifying users who heavily rely on INFORMATION_SCHEMA and transition them to ACCOUNT_USAGE. The ACCOUNT_USAGE schema holds persistent data with a slight time lag.

Benefits:

- Queries on ACCOUNT_USAGE do not utilize the cloud services layer, thereby avoiding throttling.

Factors to consider:

- There is a two- to four-hour delay between when an event occurs and when the data appears in this schema, making it unsuitable for processes requiring real-time data.

- While throttling time will be eliminated, execution times for queries on ACCOUNT_USAGE views are generally longer than those on INFORMATION_SCHEMA.

- It may not be feasible to switch some third-party tools to use ACCOUNT_USAGE.

- Access to this schema may be restricted, necessitating the resolution of any security concerns related to sharing this schema.

Create a Copy of the Data Dictionary (DD)

Establish a process that runs at regular intervals to extract the data dictionary (DD) from ACCOUNT_USAGE and ORGANIZATION_USAGE. If organizational policies restrict exposing this information to the user community, consider creating a copy of ACCOUNT_USAGE in a custom schema for users. This custom schema can include purpose-built queries and views to limit the range of accessible data and provide only the necessary information to users.

Benefits:

- Queries on the DD copy schema will not incur throttling.

- Tables in the DD copy schema can be clustered to meet user query requirements.

- Columns can be precalculated or aggregated to meet query requirements.

- Similar queries on INFORMATION_SCHEMA do not use the result cache. Queries on ACCOUNT_USAGE might use the result cache, but only if the underlying data has not changed, which is unlikely on a large or busy platform.

Factors to consider:

- There will be a lag between when an event occurs and when the data is copied into the DD copy schema. The DD copy schema load process can be set to run frequently (e.g., every 15 minutes) to minimize the lag, making it shorter than the lag for ACCOUNT_USAGE. However, there will still be a lag, which rules out using this schema for real-time processes.

- A process will be required to load the DD schema. This process will need to be designed, built, tested, scheduled, and maintained.

CHAPTER 10   CLOUD SERVICES OPTIMIZATION STRATEGIES

- There will be additional consumption requirements to load and store the DD copy schema. Some of the compute consumption may be offset by the fact that queries against the DD schema will run faster than the ACCOUNT_USAGE queries, thereby reducing warehouse uptime.

- It may not be possible to switch some third-party tools to use the DD copy schema.

Use SHOW Commands

SHOW commands can serve as an alternative to querying against INFORMATION_SCHEMA.
Benefits:

- No lag time.

- SHOW commands generally incur significantly less throttling compared to INFORMATION_SCHEMA queries.

Factors to consider:

Reengineering costs may be incurred when transitioning from INFORMATION_SCHEMA SQL statements to SHOW commands. This transition requires two steps instead of one. For example:

- One step for INFORMATION_SCHEMA queries.

    ```
    -- Use of INFORMATION_SCHEMA
    SELECT table_schema, COUNT(1) FROM information_schema.tables
    GROUP BY table_schema;
    ```

- Two steps for SHOW commands.

- One step for INFORMATION_SCHEMA queries.

    ```
    -- Use of SHOW commend
    -------------------------------------------------------------
    SHOW tables IN schema;
    SELECT "schema_name", COUNT(1) FROM table(result_scan(LAST_QUERY_ID()))
    GROUP BY "schema_name";
    ```

- SHOW commands may not provide all the required functionality or data.

- SHOW commands are not always faster than INFORMATION_SCHEMA queries, especially when filters are involved.

- Third-party tools may not be able to leverage SHOW commands.

- SHOW commands only return up to 10,000 records.

Periodic Data Dictionary Cleanup

Regularly reviewing the table and view access history, users, roles, and other objects within the database is crucial for maintaining an efficient data dictionary. Cleaning up unused and inactive objects can significantly reduce the size of the data dictionary, leading to improved performance and resource management.

For a deeper understanding of the roles and grants active in your account, consider examining the following ACCOUNT_USAGE views:

- **GRANTS_TO_ROLES:** Lists all the grants assigned to a specific role.

- **GRANTS_TO_USERS:** Displays all the grants assigned to a user.

In the section entitled "Data/Table Cleanup Based on Usage" in Chapter 7, we discuss how to identify and clean up inactive tables based on their access history. Imagine your database as a library: some tables (or books) are frequently accessed, while others gather dust. By tracing the access history, you can pinpoint which tables are rarely or not used and consider dropping them to optimize storage and metadata.

To review the grants assigned to users and roles, you can use the following query (the following image shows an example).

```
-- Review Grants assigned usage
------------------------------------------------------------
SELECT
  r1.GRANTEE_NAME AS user,
  r1.ROLE,
  r2.GRANTED_ON AS OBJECT_TYPE,
  r2.NAME AS OBJECT_NAME,
  r2.TABLE_CATALOG,
```

```
    r2.TABLE_SCHEMA,
    r2.PRIVILEGE
FROM
    SNOWFLAKE.ACCOUNT_USAGE.GRANTS_TO_USERS r1
JOIN SNOWFLAKE.ACCOUNT_USAGE.GRANTS_TO_ROLES r2
    ON r1.ROLE = r2.GRANTEE_NAME;
```

	USER	ROLE	OBJECT_TYPE	OBJECT_NAME	TABLE_CATALOG
1	SNOWFLAKE	ACCOUNTADMIN	DATABASE_ROLE	EVENTS_VIEWER	SNOWFLAKE
2	VELNATSF	ACCOUNTADMIN	DATABASE_ROLE	COST_INSIGHTS_USER	SNOWFLAKE
3	VELNATSF	ACCOUNTADMIN	DATABASE_ROLE	DATA_QUALITY_MONITORING_ADMIN	SNOWFLAKE

By regularly performing these reviews and cleanups, you can ensure that your data dictionary remains streamlined and efficient, ultimately enhancing the overall performance and governance of your Snowflake environment.

Summary

In this chapter, we explored strategies for optimizing Snowflake's data dictionary to prevent performance issues. We delved into the architecture of the cloud service layer, emphasizing the importance of managing cloud services and associated costs. Additionally, we identified factors that lead to throttling, such as high query volume and resource saturation, and provided best practices for minimizing these issues, including reducing query complexity and utilizing custom schemas. The chapter concluded with recommendations for using the ACCOUNT_USAGE schema to avoid throttling and creating a copy of the data dictionary for efficient query performance. By applying these techniques, organizations can achieve cost savings, enhance scalability, and improve overall system performance.

In the next chapter, we explore the costs associated with establishing a business continuity plan (BCP) or a disaster recovery site. We also discuss the best practices and recommendations for measuring and optimizing these cost factors.

CHAPTER 11

Data Availability Optimization Strategies

Overview

In any modern enterprise, business-critical functions must be resilient to unexpected disruptions, whether caused by human error, cyberattacks, or natural disasters. Ensuring continuity in such scenarios requires a well-defined disaster recovery (DR) and high availability (HA) strategy, combining people, processes, and technology. While DR planning can be complex, the technology layer should be as streamlined and automated as possible, since disasters are unpredictable and may occur when expert personnel are unavailable.

Snowflake addresses both HA and DR through a combination of built-in platform capabilities and configurable features:

- **High availability (HA)** is achieved through automated failover mechanisms, redundant infrastructure, and resilient services that ensure uninterrupted operations during localized failures. Snowflake multi-cluster architecture and cloud-native design allow it to automatically handle hardware failures, network issues, and service interruptions within a region.

- **Disaster recovery (DR)** is supported through cross-region and cross-cloud replication, failover groups, and automated recovery workflows. These features enable organizations to recover from large-scale outages, such as regional failures or cloud provider disruptions, with minimal intervention.

CHAPTER 11 DATA AVAILABILITY OPTIMIZATION STRATEGIES

To assess the robustness of high availability (HA) and disaster recovery (DR) strategies, two critical metrics are commonly used:

- **Recovery Point Objective (RPO)** defines the maximum acceptable amount of data loss measured in time. For example, an RPO of one hour means that up to one hour of data could be lost in a disaster. Snowflake supports asynchronous replication, allowing organizations to set practical RPO intervals based on their tolerance for data loss.

- **Recovery Time Objective (RTO)** refers to the maximum acceptable time it takes to restore normal operations after a failure. Snowflake failover groups and automated promotion of secondary environments help minimize RTO by enabling rapid recovery with minimal manual steps.

As illustrated in Figure 11-1, these capabilities enable Snowflake to design resilient, cost-effective high availability/disaster recovery (HA/DR) strategies that support customers in achieving their business continuity objectives.

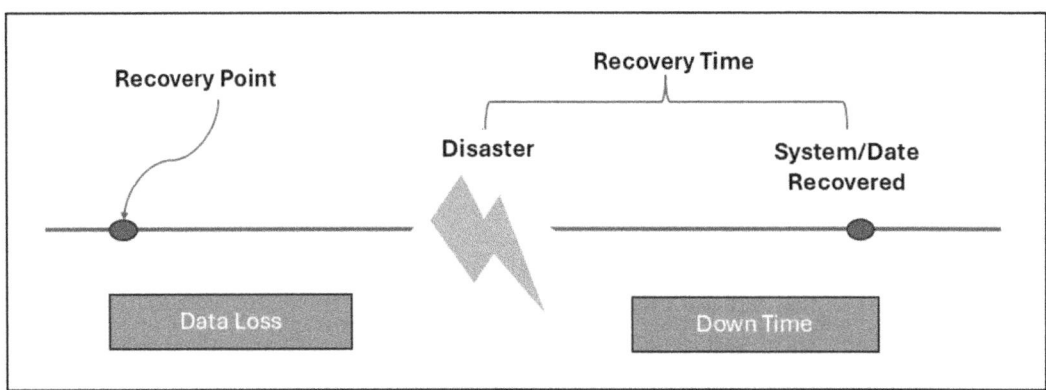

Figure 11-1. *Data availability: RPO and RTO*

With a foundational understanding of high availability (HA) and disaster recovery (DR) established, this chapter now focuses on disaster recovery replication cost-management strategies that help organizations maintain resilience while optimizing operational efficiency.

Snowflake Availability Service Level

Snowflake is architected from the ground up to be a highly available, cloud-native data platform, delivering an availability service level of 99.9% or higher across all editions. Its storage layer is built on the cloud provider's blob storage, which typically offers 11 nines (99.999999999%) of durability, ensuring exceptional data resilience.

When designing your availability strategy, it is important to align with your business recovery time objective (RTO) and recovery point objective (RPO) requirements. These expectations help determine the level of availability needed and guide the configuration of replication and failover strategies.

Table 11-1 illustrates the relationship between availability service levels and the corresponding maximum allowable downtime per year.

Table 11-1. Availability Service Level

Availability Service Level	Maximum Downtime Per Year
99%	3 days 15 hours
99.9%	8 hours 45 minutes
99.99%	52 minutes
99.999%	5 minutes

Understanding the financial implications is essential to making informed decisions about achieving these levels of availability. The next section explores the core cost components of Snowflake replication and failover, helping you balance resilience with cost-efficiency.

Cost Components of Replication and Failover

Snowflake replication and failover capabilities come with associated costs, primarily billed to the target account, the secondary or failover site. Understanding these cost drivers is essential to accurate budgeting and cost optimization.

Replication costs are influenced by four key components:

- **Data transfer:** Charges incurred when data is moved between regions or cloud providers.
- **Compute resources:** Used for managing replication processes, including background services like materialized view maintenance and search optimization.
- **Storage:** Required to maintain replicated objects and historical data.
- **Business-critical edition:** A prerequisite for enabling failover capabilities.

These costs arise from:

- The initial full replication of data.
- Incremental replication, which transfers only changed data blocks.
- Ongoing maintenance of replicated objects.

Snowflake transparently passes through cloud provider data transfer fees and uses its own compute infrastructure to handle replication tasks. When replication or refresh operations fail, Snowflake can reuse previously transferred data for up to 14 days, helping reduce redundant costs. For a detailed breakdown of applicable charges at the time of writing, refer to Table 11-2.

Table 11-2. Replication Cost Components

Data Storage	Compute	Data Transfer
Monthly data storage cost in the secondary region	Serverless compute cost to identify and copy data during replication	Cost of moving data between cloud/regions, charged by the source cloud provider
Storage pricing varies by cloud provider and region	Observed compute usage is 3-5 credits/TB of replicated data	Same cloud $20/TB Cross cloud $90-$120/TB
$20 to $40/TB per month based on cloud provider and region	Charged at the secondary region credit price	Data transfer pricing: Varies by cloud and region

After the initial seeding of the complete data from the primary system, only the changed data blocks are moved and backed up with an incremental replication.

Total Cost

Implementing replication and failover in Snowflake is essential for ensuring data resilience and operational continuity. However, without careful planning, these capabilities can lead to unnecessary costs. Organizations must adopt a strategic approach that aligns replication practices with business priorities to optimize spending while maintaining high availability. This involves monitoring usage, refining policies, and leveraging Snowflake's built-in tools to gain visibility and control.

Here are the key factors and best practices to consider when estimating and managing replication-related costs:

- Recovery point objective (RPO) and recovery time objective (RTO), as defined by business requirements
- Criticality of applications and the potential business impact of downtime
- Total volume of data (in terabytes) selected for initial replication
- Frequency of replication to the secondary region (e.g., hourly, daily)
- Percentage of data that changes and is eligible for replication
- Projected monthly or yearly data growth
- Data transfer (egress) charges and storage costs in the secondary region
- Warehouses are replicated in a suspended state initially and can be resumed when needed
- Resource monitors are replicated and continue to enforce consumption limits on the secondary account

To ensure cost efficiency while maintaining high availability, organizations should apply targeted strategies that align with their data replication goals.

Egress Data Movement Optimization

In a multi-cloud data ecosystem, managing the cost of data movement is a growing concern for organizations as they scale their data-sharing operations. To address this challenge, Snowflake recently introduced the Egress Cost Optimizer (ECO), a feature that is part of its cross-cloud auto-fulfillment framework. ECO is designed to reduce egress costs when sharing data or applications across multiple regions and cloud providers, enabling data providers to lower their service costs and improve return on investment (ROI).

As illustrated in Figure 11-2, ECO works by analyzing the configuration of a data listing, specifically the number of regions and cloud platforms involved, and then determining the most cost-efficient fulfillment strategy. When data is replicated across multiple regions, ECO routes it through a Snowflake-managed cache, avoiding redundant egress charges. For example:

- **Without ECO:** Replicating the same dataset to five regions may incur 5x egress costs.

- **With ECO:** Data is cached once, and subsequent regional replications incur zero additional egress charges.

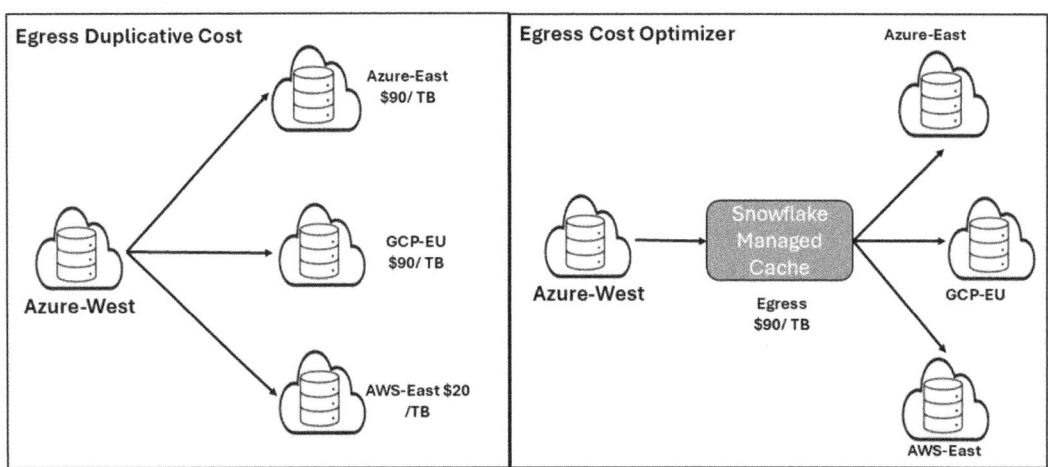

Figure 11-2. Egress cost optimizer

In contrast, when data is shared within one to two regions under the same cloud provider, ECO recognizes that additional optimization is unnecessary and bypasses the cache, ensuring cost-efficiency in all scenarios.

ECO integrates seamlessly with Snowflake existing features, including tri-secret secure (TSS) encryption, object-level replication, and scheduled listing refreshes. It introduces a cost model in which there are:

- **Initial caching charges:** One-time egress charge for full dataset replication.

- **Ongoing updates:** Only incremental changes incur egress costs.

- **Full table reloads:** Treated as new data, leading to higher costs.

The optimizer functions at the database level, meaning all listings under an account-level auto-fulfillment schedule benefit from ECO, once enabled. However, its benefits are maximized when incremental data ingestion is used, as full reloads bypass the cache and increase costs.

Ultimately, ECO offers significant savings for providers sharing data across multiple regions or clouds, while maintaining Snowflake performance, security, and operational integrity. It represents a strategic advancement in optimizing cloud data economics for scalable, cost-effective data distribution.

After looking at how to cut down on data transfer costs, the next step is to focus on managing the compute resources used during replication, another key area where costs add up quickly.

Managing Replication Spend Effectively

To address replication cost overruns, Snowflake provides various tools and configurations that allow organizations to tailor their replication strategies based on business criticality, data sensitivity, and budget constraints. As listed in Table 11-3, by understanding the cost drivers, such as data volume, replication frequency, and compute usage, teams can implement targeted optimizations that maintain resilience while minimizing spend. This overview sets the stage for exploring specific best practices that help strike the right balance between performance and cost-efficiency.

Table 11-3. Replication Optimization Focus Area

Focus Area	Optimization Goal
Monitor replication activity	Understand usage and credit consumption
Analyze database-level costs	Identify and manage high-cost replication sources
Review replication history	Detect inefficiencies and ensure reliability
Estimate future expenses	Forecast replication and failover costs
Align policies with business priorities	Match replication setup to business needs
Select object replication	Reduce compute by replicating only essential data
Plan failover testing strategically	Optimize test frequency based on workload criticality

Monitor Snowflake Replication Activity

Organizations must first gain clear visibility into how replication is being used to manage and optimize replication costs effectively in Snowflake. Snowflake provides several built-in views and table functions that allow users to monitor replication activity, track data transfer volumes, and analyze credit consumption across replication and failover groups.

The REPLICATION_GROUP_USAGE_HISTORY table function and view are key tools for this purpose. They provide detailed insights into the amount of data transferred and the credits used for replication over a specified time range. For example, to analyze usage for the current month using the account usage view, use this query:

```
SELECT start_time, end_time,
replication_group_name, credits_used, bytes_transferred
FROM snowflake.account_usage.replication_group_usage_history
WHERE start_time >= DATE_TRUNC('month', CURRENT_DATE);
```

For database-level insights, the DATABASE_REPLICATION_USAGE_HISTORY view can identify which databases are consuming the most replication credits. This query lists replicated databases and their credit usage over the last 30 days:

```
SELECT TO_DATE(start_time) AS date,
database_name, SUM(credits_used) AS credits_used
FROM snowflake.account_usage.database_replication_usage_history
```

```
WHERE start_time >= DATEADD(month, -1, CURRENT_TIMESTAMP())
GROUP BY ALL
ORDER BY 3 DESC;
```

To detect anomalies in replication trends, the following query calculates the average daily credits used per week over the past year:

```
WITH credits_by_day AS (
  SELECT start_time::date AS start_date, SUM(credits_used) AS credits_used
  FROM snowflake.account_usage.database_replication_usage_history
  WHERE start_time >= DATEADD(year, -1, CURRENT_TIMESTAMP())
  GROUP BY ALL
)
SELECT DATE_TRUNC('week', date) AS week,
       AVG(credits_used) AS avg_daily_credits
FROM credits_by_day
GROUP BY ALL
ORDER BY 1;
```

These queries empower teams to monitor replication usage proactively, identify inefficiencies, and make data-driven decisions to optimize costs. By establishing regular visibility into replication metrics, organizations can align their disaster recovery strategies with financial goals and ensure sustainable operations.

Track Database-Level Replication Costs

Understanding the cost implications of replicating individual databases is essential for optimizing Snowflake replication and failover strategy. While overall replication usage provides a high-level view, drilling down into the cost of each database allows organizations to identify high-cost contributors and make informed decisions about what to replicate and how often.

Snowflake enables this level of visibility through a combination of usage history views and cost estimation formulas. The cost for replicating a specific database can be estimated using the following formula:

Estimated Spend = (Database Bytes Transferred / Total Bytes Transferred for Replication) + (Credits Used) × (Credit Cost Rate)

This approach proportionally allocates the total replication cost based on the volume of data transferred by each database.

This example SQL script calculates the total bytes replicated for databases in a replication group over the last 30 days:

```
SELECT SUM(value:totalBytesToReplicate) AS sum_database_bytes
FROM snowflake.account_usage.replication_group_refresh_history rh,
     LATERAL FLATTEN(input => rh.refresh_stats)
WHERE rh.replication_group_name = 'SALES_ORG'
  AND rh.start_time >= CURRENT_DATE - INTERVAL '30 days';
```

To calculate the total credits used and bytes transferred for the entire replication group in the same period, use this query:

```
SELECT SUM(credits_used) AS credits_used,
       SUM(bytes_transferred) AS bytes_transferred
FROM snowflake.account_usage.replication_group_usage_history
WHERE replication_group_name = 'SALES_ORG'
  AND start_time >= CURRENT_DATE - INTERVAL '30 days';
```

By combining the results of these queries, teams can estimate the cost impact of each database and identify opportunities to reduce spend, such as excluding non-critical databases from replication or adjusting refresh frequency.

```
SELECT start_time::DATE AS start_date,
       database_name,
       SUM(credits_used) AS credits_used,
       (SUM(BYTES_TRANSFERRED))/(1024*1024*1024*1024) as TB_TRANSFERRED
FROM snowflake.account_usage.database_replication_usage_history
WHERE start_time >= CURRENT_DATE - INTERVAL '30 days'
GROUP BY ALL
ORDER BY 3 DESC;
```

Query the account usage REPLICATION_GROUP_USAGE_HISTORY view to view the credits used by the replication or failover group for account replication history over the last 30 days:

```
SELECT start_time::DATE AS start_date,
  replication_group_name,
```

```
    SUM(credits_used) AS credits_used,
    (SUM(BYTES_TRANSFERRED))/(1024*1024*1024*1024) as TB_TRANSFERRED
FROM snowflake.account_usage.replication_group_usage_history
WHERE start_time >= CURRENT_DATE - INTERVAL '30 days'
GROUP BY ALL
ORDER BY 3 DESC;
```

Audit Replication History Regularly

Regularly reviewing replication history is a critical practice for maintaining cost efficiency and operational reliability in Snowflake. By analyzing historical replication data, organizations can detect anomalies, identify inefficiencies, and ensure that replication processes are functioning as expected.

Snowflake provides several built-in views and table functions to facilitate this review. The REPLICATION_GROUP_REFRESH_HISTORY table function and view allow users to examine the history of replication events for specific replication or failover groups. This includes details such as the phase of replication, start and end times, total bytes transferred, and the number of objects replicated.

Here are some example SQL scripts.

Use this query to view the replication history of a failover group over the past seven days:

```
SELECT PHASE_NAME, START_TIME, END_TIME, TOTAL_BYTES, OBJECT_COUNT
FROM TABLE(information_schema.replication_group_refresh_history('RG_SALES'))
WHERE START_TIME >= CURRENT_DATE - INTERVAL '7 days';
```

Use this query to view the total bytes copied for a replication group where the phase is completed:

```
SELECT SUM(value:totalBytesToReplicate)
FROM
TABLE(information_schema.replication_group_refresh_history('RG_SALES'))
AS rh
WHERE rh.phase_name = 'COMPLETED'
  AND rh.start_time >= CURRENT_DATE - INTERVAL '30 days';
```

Use this query to summarize the total credits used and bytes transferred for a replication group:

```
SELECT SUM(credits_used), SUM(bytes_transferred)
FROM TABLE(information_schema.replication_group_usage_history(
  date_range_start => CURRENT_DATE - INTERVAL '30 days'));
```

To monitor the progress of ongoing replication jobs, the REPLICATION_GROUP_REFRESH_PROGRESS view or REPLICATION_GROUP_REFRESH_PROGRESS_BY_JOB table function can be used:

```
SELECT PHASE_NAME, START_TIME, END_TIME, PROGRESS, DETAILS
FROM TABLE(information_schema.replication_group_refresh_progress(
'RG_SALES'));
```

These queries provide a comprehensive view of replication activity, helping teams validate that replication is occurring as scheduled and within expected parameters. Regular audits of this data can uncover issues such as failed replications, excessive data movement, or unexpected delays, allowing for timely corrective actions.

Forecast Replication and Failover Expenses

Accurately estimating the cost of replication and failover in Snowflake is essential for proactive budgeting and cost control. Since replication involves multiple cost components, such as data transfer, compute, and storage, understanding the drivers behind these costs helps organizations avoid unexpected charges and optimize their disaster recovery strategies.

Snowflake outlines that the total spend for replication and failover is primarily influenced by two factors:

1. The amount of table data in primary databases that are part of a replication or failover group. This includes changes due to data loading or DML operations.

2. The frequency of refreshes for secondary databases or replication/failover groups.

By analyzing historical usage and refresh patterns, teams can extrapolate future costs and adjust replication schedules or scope accordingly.

Estimation Strategy

To estimate spend:

- Monitor the volume of data changes in the primary databases.
- Track how often replication refreshes occur.
- Multiply the data volume by the cost per byte transferred and the compute credits used during replication.

Although Snowflake does not provide a direct SQL script for this estimation, it suggests using the following views and functions:

- `REPLICATION_GROUP_USAGE_HISTORY`
- `REPLICATION_GROUP_REFRESH_HISTORY`

These can be queried to retrieve historical data on bytes transferred and credits used, which can then be used in cost estimation formulas.

Example (conceptual formula):

Estimated Spend = (Total Bytes Transferred) × (Data Transfer Cost Rate) + (Credits Used) × (Credit Cost Rate)

This formula can be tailored based on your Snowflake pricing agreement and cloud provider data transfer rates.

Align Replication Policies with Business Needs

Snowflake provides two powerful constructs—replication groups and failover groups—to help organizations manage data availability and disaster recovery across regions and cloud providers. Using these constructs with appropriate policies is essential for balancing resilience and cost-efficiency.

A *replication group* allows users to define:

- Which objects (e.g., databases, schemas) to replicate
- The target regions or cloud providers
- The replication frequency

A *failover group* extends this by enabling the failover of account-level objects, ensuring business continuity during outages. Each group can be configured with its own schedule and scope, allowing for granular control over replication behavior.

To optimize costs and performance:

- Include only critical objects in replication and failover groups.
- Set appropriate refresh intervals based on business needs (e.g., every 30 minutes, hourly).
- Avoid unnecessary replication to regions where failover is not required.

Additionally, data transfer costs vary by cloud provider and region, so policies should consider geographic pricing differences.

Best Practice Tip In the event of a regional outage, Snowflake recommends promoting replicated secondary account objects and databases to read-write primary objects to resume operations quickly.

Here is a conceptual example of how you might define a replication group:

```
CREATE REPLICATION GROUP demo_replication_group
  OBJECTS (DATABASE demo_db, SCHEMA demo_schm)
  TO REGION 'us-west-2'
  REFRESH EVERY 60 MINUTES;
```

And for failover:

```
CREATE FAILOVER GROUP demo_failover_group
  OBJECTS (DATABASE demo_db, USER demo_schm)
  TO REGION 'us-east-1';
```

These constructs should be aligned with your disaster recovery policies, data criticality, and budget constraints.

CHAPTER 11   DATA AVAILABILITY OPTIMIZATION STRATEGIES

Optimize Object Selection for Replication

One of the most effective ways to control replication and failover costs in Snowflake is to replicate only the essential objects selectively. Not all data or objects require high availability across regions or cloud providers. As illustrated in Figure 11-3, narrowing the replication scope to mission-critical assets can lead to substantial savings in data transfer, compute, and storage expenses.

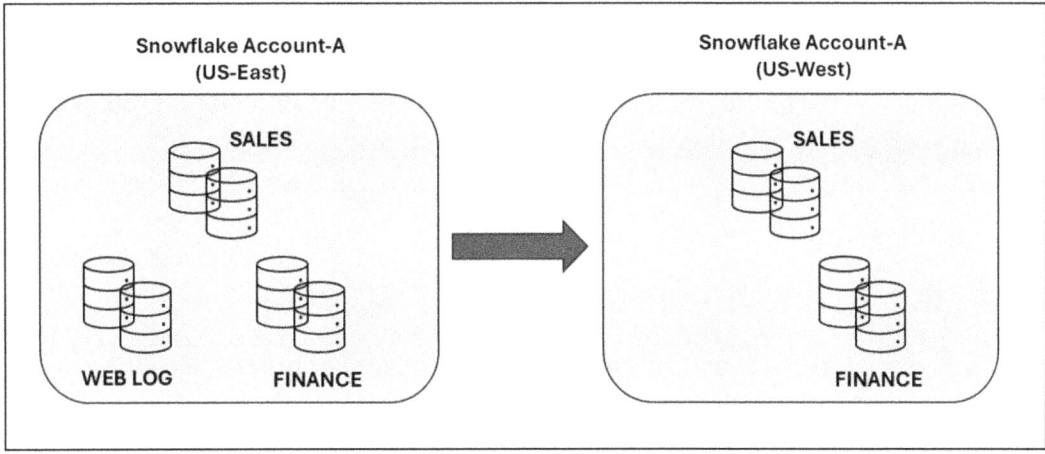

Figure 11-3. Optimize object selection for replication

Snowflake allows users to define replication and failover groups with fine-grained control over:

- Which objects (e.g., databases, schemas, users) are included
- Where they are replicated (target region or cloud)
- How often they are refreshed

This flexibility enables teams to align replication strategies with disaster recovery priorities and cost management goals.

Best practices:

- Replicate only mission-critical databases required for failover.
- Exclude non-essential or archival data from replication groups.
- Adjust refresh frequency based on data volatility and business needs (e.g., every 30 minutes vs. hourly).

Example configuration:

```
CREATE REPLICATION GROUP critical_data_group
  OBJECTS (DATABASE sales_data, SCHEMA finance.reports)
  TO REGION 'us-west-2'
  REFRESH EVERY 60 MINUTES;
```

This example replicates only the `sales_data` database and `finance.reports` schema to a secondary region every hour, balancing availability with cost.

Tip Snowflake supports setting different refresh intervals for different groups, allowing you to optimize replication frequency per use case.

Set Failover Testing Frequency Strategically

Establishing a regular cadence for test failovers is a critical component of a resilient disaster recovery strategy in Snowflake. These planned failovers simulate real-world outages, allowing organizations to validate their readiness, measure recovery time objectives (RTO), and ensure that failover configurations function as expected.

The frequency of test failovers should be tailored to the criticality of workloads:

- High-impact systems (e.g., financial reporting, customer-facing apps) may require monthly or quarterly failovers.
- Lower-priority systems may be tested semi-annually or annually.

Snowflake also supports chaos engineering practices, where faults are intentionally injected during "game days" to simulate disaster scenarios. These exercises help teams:

- Test hypotheses (e.g., "If region X fails, query failure rate should remain <10%")
- Monitor system behavior under stress
- Improve operational response and automation

Best practices:

- Automate test failovers using scripts or orchestration tools.
- Monitor metrics such as query success rate, replication lag, and recovery time.
- Document outcomes and refine your failover strategy accordingly.

As an example, this command simulates a failover by promoting the secondary environment to active status, allowing teams to validate application continuity and data integrity.

```
-- Promote secondary objects to primary in a failover group
ALTER FAILOVER GROUP demo_failover_group PROMOTE;
```

Replication compute costs can escalate quickly as data grows and refreshes become more frequent. Without active monitoring and smart policies, organizations risk overspending. Understanding key cost drivers helps teams optimize usage while maintaining high availability.

Next, while compute and data transfer are major cost drivers, storage in the secondary account is another critical factor that can significantly impact overall replication expenses.

Storage Consumption of the Secondary Account

When replicating data in Snowflake, especially across regions or accounts, it is important to recognize that storage costs are incurred independently on the secondary account. This means that even though the data originates from a primary account, the replicated copy in the secondary account is treated as a separate storage entity for billing purposes.

What Drives Storage Costs on the Secondary Account?

The following aspects increase storage costs on the secondary account:

Replicated data volumes: Every time a database or failover group is replicated, the full dataset (or the delta during refreshes) is stored in the secondary account. This includes:

- Table data

- Metadata
- Materialized views
- Search optimization structures

Ongoing storage charges: Snowflake applies standard storage rates to the data stored in the secondary account, just as it does in the primary. This means:

- The more data you replicate, the higher the storage cost.
- Storage is billed monthly based on the average daily volume stored.

Background services: In addition to raw data, automatic background processes, such as those maintaining materialized views or search optimization, also consume storage and compute resources. These are billed under Snowflake serverless features and can add to the overall cost footprint of the secondary account.

Key Considerations for Managing Storage Costs

To keep storage consumption, and associated costs, under control in the secondary account, consider the following strategies:

- Replicate only the databases, schemas, or objects that are essential for business continuity or compliance. Avoid replicating non-critical or archival data.
- Set refresh intervals based on data volatility and business needs. Adjusting the refresh frequency can increase storage usage due to versioning and change tracking.
- Monitor usage regularly using built-in views such as ACCOUNT_USAGE. REPLICATION_GROUP_USAGE_HISTORY to track storage consumption and identify optimization opportunities.

Storage consumption on the secondary account is a direct contributor to overall replication costs. Without careful planning, organizations may incur unnecessary expenses from storing redundant or infrequently used data. By understanding what drives these costs and applying targeted controls, teams can maintain high availability while optimizing their Snowflake spend.

Beyond disaster recovery, replication capabilities also support broader strategic initiatives. One such use case is account migration, where replication plays a key role in enabling seamless transitions across regions or cloud platforms.

Leverage Replication for Strategic Account Migration

Snowflake replication and failover capabilities are not only essential for disaster recovery, they also serve as powerful tools for account migration during mergers and acquisitions (M&A) or strategic cloud transitions. These scenarios often require moving data and account configurations across regions or cloud providers with minimal disruption to business operations.

By using replication groups and failover groups, organizations can:

- Seamlessly replicate databases, schemas, and account-level objects to a new region or cloud platform.

- Maintain point-in-time consistency across environments.

- Promote the replicated environment to primary once the migration is complete.

This approach ensures business continuity, reduces downtime, and avoids the complexity of manual data exports and reconfigurations.

Use cases:

- **M&A integration:** Consolidate Snowflake accounts from different business units or acquired entities.

- **Cloud strategy shift:** Transition from a single-cloud to a multi-cloud or cross-region architecture.

Example:

```
-- Step 1: Create a failover group to replicate account objects
CREATE FAILOVER GROUP sales_migrate_group
  OBJECTS (DATABASE sales_data, USER analytics_user)
  TO REGION 'us-east-1';
```

```
-- Step 2: Promote the replicated environment to primary
ALTER FAILOVER GROUP sales_migrate_group PROMOTE;
```

This script outlines a basic migration flow: replicate critical objects to a new region, then promote them to active status once validated.

Best practices:

- Test the migration in a staging environment before production cutover.

- Monitor replication lag and validate data integrity post-promotion.

- Coordinate with stakeholders to minimize the impact during the switch.

Conceptual Example: Estimating Replication Costs

To illustrate how replication costs can be estimated in Snowflake, this section walks through a conceptual example using AWS as the cloud provider. This scenario reflects a common enterprise setup in which a company replicates data from its U.S.-based Snowflake account to a European region to support local compliance and improve query performance for their EU-based analysts.

Scenario overview:

- The team replicates half of their 65 TB dataset (about 32.5 TB) to the EU region.

- Replication jobs run daily, using a Medium Standard virtual warehouse for four hours per day.

- The replicated data is stored in the EU region and updated regularly.

- Data is transferred across regions, incurring egress charges.

The following section breaks down the cost implications of this setup.

CHAPTER 11 DATA AVAILABILITY OPTIMIZATION STRATEGIES

Compute: Running the Replication Jobs

To move and transform data, the team uses a Medium warehouse (four credits/hour). Running it four hours a day for a month:

- **Credits used:** 4 credits/hour × 4 hours/day × 30 days = 480 credits
- **Cost (@ $2/credit):** 480 × $2 = $960

This covers the compute power needed to extract, prepare, and write data to the EU region.

Storage: Holding the Replicated Data

The replicated 32.5 TB of data is stored in the EU region:

- **Storage cost (@ $23/TB):** 32.5 TB × $23 = $747.50/month

This is the ongoing cost of maintaining a copy of the data in the destination region.

Data Transfer: Moving Data Across Regions

Since the team is using AWS as their cloud provider, Snowflake charges for data egress when transferring data between AWS regions (e.g., from US East to EU Central). These inter-region transfers are billed per terabyte.

- **Transfer volume:** 32.5 TB/month
- **Egress rate:** $20 per TB (typical inter-region rate)
- **Transfer cost:** 32.5 × $20 = $650

This cost reflects the outbound data transfer from the U.S.-based Snowflake account to the EU region on AWS infrastructure.

Total Monthly Replication Cost

As shown in Table 11-4, the following summary consolidates all components to provide an estimate of the total monthly cost associated with replicating data in Snowflake under the defined scenario.

Table 11-4. Total Monthly Replication Cost

Cost Component	Monthly Cost
Compute	$960
Storage	$747.50
Data Transfer	$650
Total	$2,357.50

This conceptual example helps illustrate how replication in Snowflake, while powerful, involves a combination of compute, storage, and data transfer costs. By understanding these components, teams can better plan and optimize their cloud data strategies.

Summary

This chapter explained how Snowflake helps keep your data available and protected during outages by using features like replication and failover. While these tools are great for making sure your systems stay up and running, they can also lead to extra costs, especially in the backup (secondary) account, where storage, compute, and data transfer add up. The chapter provided practical tips on how to keep these costs under control, like only copying important data, adjusting how often you replicate, and using built-in tools to track and predict spending. It is all about staying prepared without overspending.

In the next chapter, the focus shifts to optimizing the Snowflake cloud services layer. You learn how to identify patterns that drive cloud services usage, manage metadata operations more efficiently, and reduce unnecessary overhead. These strategies help improve performance and control costs even further, building on the foundation of resilient and cost-effective data availability.

CHAPTER 12

Snowflake AI/ML Cost Management

Overview

Snowflake features two broad categories of intelligent tools built on artificial intelligence (AI) and machine learning (ML). These tools empower users to extract insights and build intelligent applications directly within the Snowflake Data Cloud. These capabilities are powered by Snowflake-managed compute resources and are billed using Snowflake credits.

Among these offerings, large language model (LLM) functions are a key component, with costs primarily determined by the number of tokens processed, where a token represents a small unit of text, typically around four characters.

Snowflake AI and ML capabilities can be broadly categorized into two groups:

- **Cortex AI-specific features:** These are built on top of LLMs and are designed for unstructured data processing, natural language understanding, and generative AI tasks.

- **Snowflake ML platform features:** These include tools for building, training, and deploying custom ML models, as well as features that enhance explainability and integration into applications.

In the following sections, we explore each category in detail, along with their associated cost considerations to help you manage and optimize your Snowflake AI usage effectively.

Snowflake AI Features Credit Table

Snowflake Cortex AI offers a powerful suite of AI capabilities designed to bring LLMs and intelligent automation directly into the Snowflake Data Cloud, as shown in Figure 12-1. It also introduces *AISQL*, a feature that enables business analysts and non-programmers to leverage AI through prompt-based SQL, making advanced capabilities accessible without the need for coding expertise.

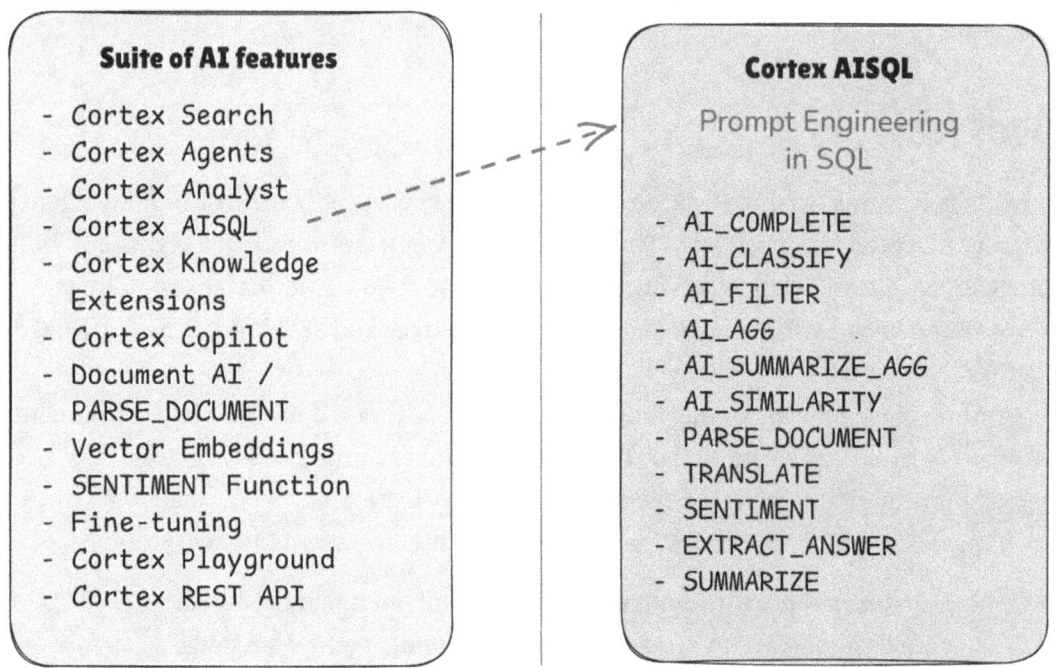

Figure 12-1. *Snowflake AI/AISQL features*

When it comes to AI pricing, Snowflake provides transparency into credit consumption, helping users better understand and manage associated costs. These features are billed according to the pricing structure outlined in Table 12-1 of the official Credit Consumption Table on the number of tokens processed.

In Tables 12-1a, 12-1b, and 12-1c, the term "token" refers to the smallest unit of text processed by the underlying model (in this case, the AI model). The conversion of raw input or output text into tokens depends on the specific Cortex model. Billing is based on the number of tokens processed, which, depending on the model, may include only input tokens or both input and output tokens.

Table 12-1a. *Snowflake AI Features Credit Tables, Cortex AI Functions (Source Courtesy Snowflake Inc.)*

Cortex Feature	Snowflake-managed compute (Credits per one million Tokens)
AI Complete – claude-3-5-sonnet	2.55
AI Complete – claude-3-7-sonnet	2.55
AI Complete – claude-4-sonnet	2.55
AI Complete – claude-4-opus	12.00
AI Complete – deepseek-r1	1.03
AI Complete – llama3.1-405b	3.00
AI Complete – llama3.1-70b	1.21
AI Complete – llama3.1-8b	0.19
AI Complete – llama3.3-70b	1.21
AI Complete – llama4-maverick	0.25
AI Complete – llama4-scout	0.14

Note Refer to the Snowflake Credit Consumption Table at www.snowflake.com/legal-files/CreditConsumptionTable.pdf for details on credit usage for all supported Cortex AI functions.

Table 12-1b. *Snowflake AI Features Credit Table, Fine Tuning (Source Courtesy Snowflake Inc.)*

Feature	Snowflake-managed compute (Credits per one million Tokens)	
	Training	Cortex Complete (Inference)
Cortex Fine-tuning – llama3.1-70b	3.40	2.42
Cortex Fine-tuning – llama3.1-8b	0.64	0.38
Cortex Fine-tuning – mistral-7b	0.64	0.24
Cortex Fine-tuning – mixtral-8x7b	3.40	0.44

CHAPTER 12 SNOWFLAKE AI/ML COST MANAGEMENT

Table 12-1c. *Snowflake AI Features Credit Table, Other Services (Source Courtesy Snowflake Inc.)*

Table 6(c): Snowflake AI Features Credit Table, Other	
Feature	Snowflake-managed compute
Cortex Analyst	67 Credits per 1,000 messages[16]
Cortex Search	6.3 Credits per GB/mo of indexed data
Document AI	8 Credits per hour of compute
Parse Document – Layout	3.33 Credits per 1,000 pages
Parse Document – OCR	0.5 Credits per 1,000 pages

For further understanding, see Figure 12-2.

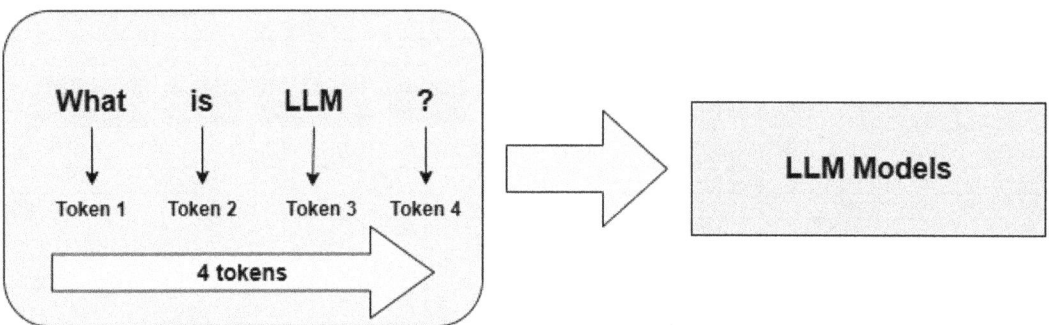

Figure 12-2. *Tokens processed by an LLM model*

Billing for Snowflake AI features may be based on:

- **Token usage:** Charges are calculated based on the total number of tokens processed. This may include input tokens only, or both input and output tokens, depending on the feature.

- **Compute time:** As specified in Table 12-1c, some features are billed based on the compute time required to complete a task. This compute time may be normalized depending on the type of compute resources utilized.

Understanding these billing mechanisms is essential for managing and optimizing Cortex AI costs within the Snowflake ecosystem.

To begin, the next section delves into Cortex AI-specific features, highlighting their pricing structure, which is a critical aspect of managing AI-driven tasks effectively.

Cortex AI Features and Cost Considerations

Snowflake Cortex AI features are purpose-built to handle unstructured data, enable natural language interactions, and support generative AI use cases, all without requiring users to move data outside the platform.

Cortex AI is structured around modular components that serve different roles in the AI workflow. These include intelligent agents, natural language interfaces, document processing tools, and hybrid search capabilities. Each feature is tightly integrated with Snowflake compute and storage infrastructure, ensuring scalability, security, and governance.

The suite of key services includes the features outlined in Table 12-2.

Table 12-2. Cortex AI Features

Cortex AI Feature	Use Case
Cortex Agents	Intelligent agents that automate workflows and interact with users using natural language.
Cortex Analyst	Enables natural language querying and analysis of structured data using LLMs.
Cortex Knowledge Extensions	Allows integration of external knowledge sources into AI workflows.
Cortex Search	Hybrid (vector + keyword) search engine for low-latency, high-quality search over Snowflake data. Ideal for RAG applications.
Cortex AISQL	SQL extensions for invoking LLMs and AI functions directly in SQL queries.
Cortex Playground	Interactive environment to experiment with AI functions and build prototypes.
Cortex REST API	Programmatic access to Cortex features for integration into external applications.

(*continued*)

Table 12-2. (*continued*)

Cortex AI Feature	Use Case
Cortex Copilot	AI assistant embedded in the Snowflake UI to help users write SQL, explore data, and build apps.
Document AI	Extracts structured data from unstructured documents (PDFs, images, etc.).
Cortex PARSE_DOCUMENT	Parses and extracts structured content from documents using AI.
Vector Embeddings	Converts text into vector representations for semantic search and ML.
Sentiment (SNOWFLAKE.CORTEX)	Built-in sentiment analysis function.
Fine-tuning	Customizes LLMs with your own data for domain-specific tasks.

AI Cost Considerations

The cost structure for Cortex AI features follows a modular, usage-based pricing model, with costs tied to compute, storage, and token consumption. This structure supports both experimentation and production-scale deployments. As listed in Table 12-3, understanding these cost considerations is essential to managing AI workloads efficiently.

Table 12-3. *AI Cost Considerations*

Cortex AI Feature	Cost Considerations
Cortex Agents	Incur costs through the Cortex Search and Cortex Analyst services they orchestrate. These include compute for embedding, indexing, and querying.
Cortex Analyst	Uses LLMs to generate SQL from natural language. Costs are tied to LLM usage and query execution on Snowflake compute. Fully managed service, so pricing reflects convenience and scalability.

(*continued*)

Table 12-3. (*continued*)

Cortex AI Feature	Cost Considerations
Cortex Knowledge Extensions (CKEs)	Built on Cortex Search, so they inherit its cost model; embedding tokens and serving compute and storage.
Snowflake Copilot	Currently available free of charge, providing an excellent opportunity to explore various LLM features.
Cortex Search	Costs include: • Token embedding: Charged per token embedded when indexing text data. • LLM usage: Only applicable when used in RAG or chat scenarios with Cortex LLM functions; billed separately. • Serving compute: Billed per GB/month of indexed data. • Index refreshes: Included in serving compute; no separate cost.
Document AI	Costs are based on document processing volume and LLM usage. Charges may include: • Document ingestion and parsing: Based on the number and size of documents. • LLM inference: For extracting structured data from unstructured documents. • Integration with Cortex search: May incur additional embedding and indexing costs if used in conjunction.

With a clear understanding of Cortex AI core features and their associated cost structures, we now turn to one of the most powerful and flexible components: AISQL functions. These functions enable direct interaction with LLMs using SQL and Python, making advanced AI capabilities accessible to analysts, engineers, and developers.

AISQL Functions and Cost Considerations

The Cortex AISQL pricing structure is primarily based on token usage, offering flexibility and scalability tailored to diverse AI-driven analytics needs. Cost management is influenced by selecting the appropriate models and their specific parameters, as outlined in Table 12-4.

Table 12-4. AISQL Cost Considerations

AISQL Function	Use Case	Example Models	Cost Considerations
AI_COMPLETE	Generates completions for text or images using a selected LLM.	snowflake-arctic, llama3.2-1b, gemma-7b	Charged per token (input + output). Structured outputs and metadata retrieval do not add extra cost.
AI_CLASSIFY	Classifies text or images into user-defined categories. Supports multi-label and image classification.	claude-3-5-sonnet, llama3-70b	Token-based cost, including category descriptions and examples.
AI_FILTER	Returns a Boolean (True/False) for filtering based on text or image input.	claude-3-5-sonnet, mistral-large	Token-based cost, depending on input complexity.
AI_AGG	Aggregates text across rows to extract insights using a prompt.	claude-3-5-sonnet, reka-core	Token-based cost across all rows processed. Not limited by context window.
AI_SUMMARIZE_AGG	Summarizes text across multiple rows.	claude-3-5-sonnet, mistral-large2	Token-based cost across all rows. Not limited by context window.
AI_SIMILARITY	Computes embedding similarity between two inputs.	snowflake-arctic, llama3.1-8b	Embedding cost per input; similarity computation is lightweight.

(continued)

CHAPTER 12   SNOWFLAKE AI/ML COST MANAGEMENT

Table 12-4. (*continued*)

AISQL Function	Use Case	Example Models	Cost Considerations
PARSE_DOCUMENT	Extracts text or layout from documents using OCR or layout mode.	claude-3-5-sonnet (for layout interpretation)	Charged per document processed; layout mode may incur higher cost.
TRANSLATE	Translates text between supported languages.	mistral-large, gemma-7b	Token-based cost depending on input and output language length.
SENTIMENT	Extracts sentiment scores from text.	claude-3-5-sonnet, jamba-1.5-mini	Token-based cost depending on input length.
EXTRACT_ANSWER	Extracts answers from unstructured data.	claude-3-5-sonnet, llama3.3-70b	Token-based cost; depends on input size and complexity.
SUMMARIZE	Summarizes a given text.	claude-3-5-sonnet, openai-o4-mini	Token-based cost depending on input length.

In addition to AISQL capabilities, Snowflake expands its offerings with a robust machine learning platform, where pricing plays a critical role. The platform cost structure is tailored to specific ML operations, ensuring clarity and flexibility for diverse needs.

Optimizing Token Usage in Cortex AI

Tokens are units of text, typically consisting of about four characters, and they serve as the fundamental currency of LLMs. Both input and output tokens contribute to usage costs, making efficient token management essential for controlling expenses.

Advanced models such as Claude, GPT-4, and Mistral offer exceptional performance. However, without careful usage, they can lead to high operational costs. To maintain a balance between performance and budget, it is important to implement strategies that optimize token consumption, as outlined in Table 12-5.

Table 12-5. Token Usage Optimization Strategies

Optimization	Strategy	Why It Saves Tokens
Use smaller models when possible	Choose lightweight models like llama3.2-1b or gemma-7b for simple tasks.	Smaller models often require fewer tokens and are cheaper per token.
Limit output tokens	Set a max_tokens parameter in functions like AI_COMPLETE.	Prevents unnecessarily long responses that increase cost.
Trim input text	Preprocess and remove irrelevant or redundant content before sending it to the model.	Reduces input token count directly.
Use the COUNT_TOKENS function	Estimate token usage before running a query.	Helps avoid exceeding limits and allows for better cost planning.
Avoid repetition in prompts	Do not repeat instructions or context in every row when using batch processing.	Reduces token duplication across rows.
Use TRY_COMPLETE for fault tolerance	Avoid retries and errors that may consume tokens.	Saves cost by returning NULL instead of failing.
Batch processes with AI_AGG or AI_SUMMARIZE_AGG	Aggregate multiple rows into one prompt when possible.	Reduces overhead from repeated prompts.
Use shorter prompts	Be concise and specific in your instructions.	Every word counts—literally.
Avoid overly large context windows	Do not use high-context models unless needed (e.g., 200K token models).	Smaller context windows are more cost-effective for short tasks.

Token optimization is not just about reducing cost; it is about efficiently leveraging model capabilities. Here is the reasoning behind the strategies:

> **Model selection matters:** Larger models like Claude or GPT-4 offer broader reasoning and language capabilities, but they also consume more tokens and cost more per token. For simpler tasks (e.g., keyword extraction, basic classification), smaller models can perform just as well at a fraction of the cost.

Token limits and context windows: Each model has a maximum context window (e.g., 128K or 200K tokens). Sending excessive input wastes capacity and may reduce the space available for meaningful output. Keeping inputs concise ensures that the model focuses on relevant content and avoids truncation.

Prompt engineering efficiency: Prompts are part of the input token count. Repetitive or verbose prompts inflate costs. Well-crafted, minimal prompts reduce token usage while maintaining clarity. This is especially important in batch operations where the same prompt is applied across many rows.

Output control: Without a max_tokens limit, models may generate unnecessarily long responses. This not only increases cost but can also introduce irrelevant or verbose content. Setting output limits ensures responses are focused and cost-effective.

Batching and aggregation: Functions like AI_AGG and AI_SUMMARIZE_AGG allow you to process multiple rows in a single prompt. This reduces the overhead of repeated instructions and leverages the model's ability to generalize across data, improving both efficiency and insight.

Preprocessing and filtering: Removing irrelevant text, HTML tags, or boilerplate before sending data to the model reduces token count and improves model performance. Clean input means efficient output.

Cost awareness tools: Functions like COUNT_TOKENS and TRY_COMPLETE are designed to help you estimate and manage token usage. They act as safeguards against unexpected costs and failed executions.

By implementing these strategies, organizations can optimize token usage while keeping costs in check. To monitor and analyze AI service costs effectively, Snowflake provides powerful tools and views for detailed tracking.

CHAPTER 12   SNOWFLAKE AI/ML COST MANAGEMENT

Comparing Token Usage Across Cortex AI Models

When working with Cortex AI, choosing the right model size is essential for balancing performance, cost, and compliance. The Snowflake platform supports a wide range of models, from lightweight to heavyweight, enabling users to match model complexity to the task at hand.

> **Note** The following comparison is for illustration purposes only. Actual token usage may vary depending on the prompt structure, model version, and other contextual factors.

To illustrate how token usage varies across different models, we use the COUNT_TOKENS helper function to evaluate a sample prompt.

Prompt: "Is Snowflake built on cloud MPP architecture?"

```
SELECT SNOWFLAKE.CORTEX.COUNT_TOKENS('snowflake-arctic', 'Is Snowflake built on cloud MPP architecture?');
```

Token Usage Comparison

Table 12-6 shows a comparison of token usage across different models for the same prompt.

Table 12-6. Cortex Model Token Usage

Model Name	Tokens Used
snowflake-arctic	13
llama3.2-1b	10
gemma-7b	9
llama4-maverick	11
snowflake-llama-3.1-405b	11
llama4-scout	11

This data shows that Gemma-7b and Llama 3.2-1b are the most token-efficient, using up to 30% fewer tokens than the other models.

Impact of Guardrails on Token Usage

Snowflake Cortex allows users to enable guardrails, which apply safety, compliance, and ethical filters to model outputs. While these guardrails enhance trust and governance, they also introduce additional token overhead.

Table 12-7 shows a comparison of token usage for Llama 3.2-1b with and without guardrails.

Table 12-7. Guardrails Token Usage

Model	Guardrails	Prompt Tokens	Completion Tokens	Guardrails Tokens	Total Tokens
llama3.2-1b	Disabled	20	100	0	120
llama3.2-1b	Enabled	20	24	258	302

With guardrails enabled, the total token usage more than doubles due to the additional safety processing. This highlights a key tradeoff: guardrails improve safety and compliance but increase token consumption.

By choosing smaller models for simpler tasks and enabling guardrails where necessary, organizations can balance efficiency, safety, and performance.

Monitoring AI Services Credit Usage

To monitor credit consumption for AI services, including LLM functions, use the Snowflake's account and organization usage views. These views offer detailed information on daily credit usage across various services.

Daily AI Services Credit Usage

To track credits used for AI services, including LLM functions, use the METERING_DAILY_HISTORY view. This query shows how many Snowflake credits were billed each day for AI services. It helps track overall AI-related spend overtime.

```
SELECT
    USAGE_DATE,
    SERVICE_TYPE,
    ROUND(SUM(CREDITS_BILLED), 2) AS CREDITS_BILLED
```

```
FROM SNOWFLAKE.ACCOUNT_USAGE.METERING_DAILY_HISTORY
WHERE SERVICE_TYPE = 'AI_SERVICES'
GROUP BY ALL
ORDER BY USAGE_DATE;
```

USAGE_DATE	SERVICE_TYPE	CREDITS_BILLED
2025-06-15	AI_SERVICES	120.00
2025-06-16	AI_SERVICES	112.00

Detailed LLM Function Usage

This query returns detailed logs of every LLM function call, including the model name, the function used, the token counts, and the credits consumed. It's useful for granular auditing.

```
SELECT START_TIME,FUNCTION_NAME,MODEL_NAME,TOKEN_CREDITS
FROM SNOWFLAKE.ACCOUNT_USAGE.CORTEX_FUNCTIONS_USAGE_HISTORY;
```

START_TIME	FUNCTION_NAME	MODEL_NAME	TOKEN_CREDITS
2025-06-15 10:00:00.000 -0700	EMBED_TEXT	snowflake-arctic-embed-l-v2.0	0.000003100
2025-06-16 12:00:00.000 -0700	COMPLETE	snowflake-llama-3.3-70b	0.000168200
2025-06-16 12:00:00.000 -0700	COMPLETE	claude-4-sonnet	0.000833850
2025-06-16 12:00:00.000 -0700	COMPLETE	snowflake-arctic	0.000352800
2025-06-16 12:00:00.000 -0700	COMPLETE	gemma-7b	0.000024360
2025-06-16 12:00:00.000 -0700	COMPLETE	snowflake-arctic	0.000089040
2025-06-16 12:00:00.000 -0700	COMPLETE	llama3.2-1b	0.000037960

Credit and Token Consumption by Query

To analyze credit and token usage for queries that utilize AI models within your account, see the following query. This query allows you to track resource consumption at the query level for better cost management and optimization.

```
SELECT QUERY_ID,MODEL_NAME,FUNCTION_NAME,TOKENS, TOKEN_CREDITS
FROM SNOWFLAKE.ACCOUNT_USAGE.CORTEX_FUNCTIONS_QUERY_USAGE_HISTORY;
```

QUERY_ID	MODEL_NAME	FUNCTION_NAME	TOKENS	TOKEN_CREDITS
01bd0cec-0002-fe97-0008-95b20004107a	snowflake-arctic-embed-l-v2.0	EMBED_TEXT	62	0.000003100
01bd12ee-0003-0382-0008-95b20005273a	snowflake-arctic	COMPLETE	169	0.000141960
01bd12f0-0002-fe97-0008-95b200041716	claude-4-sonnet	COMPLETE	327	0.000833850
01bd12f1-0003-03d5-0008-95b20005169a	llama3.2-1b	COMPLETE	785	0.000031400
01bd12f1-0003-03d5-0008-95b20005169e	openai-gpt-4.1	COMPLETE	375	0.000525000
01bd12f2-0003-033d-0008-95b2000544d6	snowflake-llama-3.3-70b	COMPLETE	580	0.000168200
01bd12f6-0003-033d-0008-95b2000544ee	snowflake-arctic	COMPLETE	106	0.000089040

Daily Token Usage by Model

With this query, you can track the number of tokens processed by each model on a daily basis to better understand usage patterns. This information helps correlate model activity with cost trends over time.

```
SELECT
    START_TIME::DATE AS USAGE_DATE,
    MODEL_NAME,
    (SUM(TOKEN_CREDITS)) AS CREDITS_USED
FROM SNOWFLAKE.ACCOUNT_USAGE.CORTEX_FUNCTIONS_USAGE_HISTORY
WHERE LENGTH(MODEL_NAME) > 1
GROUP BY ALL
ORDER BY USAGE_DATE, CREDITS_USED DESC;
```

USAGE_DATE	MODEL_NAME	CREDITS_USED
2025-06-15	snowflake-arctic-embed-l-v2.0	0.000003100
2025-06-16	openai-gpt-4.1	0.000907200
2025-06-16	claude-4-sonnet	0.000833850
2025-06-16	claude-3-5-sonnet	0.000772650
2025-06-16	snowflake-arctic	0.000441840
2025-06-16	snowflake-llama-3.3-70b	0.000168200
2025-06-16	llama3.2-1b	0.000037960

By examining these metrics and understanding their implications, users can gain deeper insights into the cost structure and token consumption of various AI functions, enabling more informed decision-making.

Snowflake ML Capabilities and Cost Consideration

Beyond Cortex AISQL, the platform offers a broader suite of machine learning capabilities through Snowpark ML and the Native App Framework. These tools support model training, explainability, and seamless integration of ML into applications. Pricing is generally based on the computational resources consumed during training and inference, rather than token-based billing, as outlined in Table 12-8.

Table 12-8. ML Functions and Cost Management

Function	Use Case	Cost Management Tips
FORECAST	Predicts future values of a metric based on time-series data.	Use smaller time windows or sample data to reduce compute time during training.
ANOMALY_DETECTION	Identifies outliers in time-series data.	Run on aggregated data where possible.
CLASSIFICATION	Sorts rows into categories based on predictive features.	Limit feature sets to essential variables; use stratified sampling for training datasets.
TOP_INSIGHTS	Identifies surprising or impactful dimensions affecting a metric.	Apply filtered datasets or specific segments to minimize unnecessary scans.

For tasks that demand high computational power, such as deep learning, large-scale model training, or real-time inference, GPU-enabled compute is available through Snowpark Container Services. These resources unlock significant performance gains but require thoughtful planning to ensure efficiency and cost-effectiveness.

Building Cost-Effective ML Models

As ML models grow in complexity and data volumes continue to scale, the demand for efficient training infrastructure becomes critical. This challenge is addressed through GPU-accelerated model training enabled directly within the platform using Notebooks and Snowpark Container Services. This integration allows data scientists to train models where their data resides, eliminating the friction of external orchestration or infrastructure management.

Why GPU Acceleration Matters

GPUs are designed for parallel computation, making them ideal for training deep learning models and accelerating traditional ML algorithms like XGBoost. Leveraging GPUs can reduce training time from hours to minutes, enabling faster experimentation and iteration.

Snowflake Notebooks and Container Services

Notebooks provide a native development environment for Python-based ML workflows. When paired with Container Services, users can select GPU-backed environments with pre-installed ML libraries such as PyTorch, TensorFlow, and XGBoost. This eliminates the need for manual environment setup or dependency management.

Best Practices for GPU Optimization

- **Minimize data transfer:** Keep data in Snowflake and use Snowpark or Snowflake datasets to avoid unnecessary I/O.
- **Use efficient data formats:** Convert data to NumPy arrays or arrow tables before feeding into models.
- **Profile your workload:** Use Snowflake's observability tools to monitor GPU usage and identify bottlenecks.

Cost Considerations

While GPU acceleration offers performance benefits, it also comes with cost implications:

- Compute charges are based on the type and duration of GPU usage.
- Container runtime usage is billed separately from the standard Snowflake compute.

- To utilize the GPU efficiently:
  - Optimize the idle timeout setting for lower environment.
  - Users can explicitly end sessions to release the GPU, promoting more cost-effective usage.

Note GPU compute is significantly more expensive than the standard Snowflake compute due to its specialized hardware and high-performance infrastructure. Refer to the Snowflake Credit Consumption Table at www.snowflake.com/legal-files/CreditConsumptionTable.pdf for additional information.

Monitor Snowpark Container Service Usage

Snowflake offers built-in mechanisms to monitor usage at both the account and organizational level. These tools provide visibility into how compute pools are utilized over time, enabling teams to track consumption patterns, allocate resources efficiently, and maintain budgetary control. This level of transparency is especially valuable in environments where multiple teams or applications share infrastructure.

Use the built-in NVIDIA SMI command to monitor GPU usage, as shown in Figure 12-3.

CHAPTER 12 SNOWFLAKE AI/ML COST MANAGEMENT

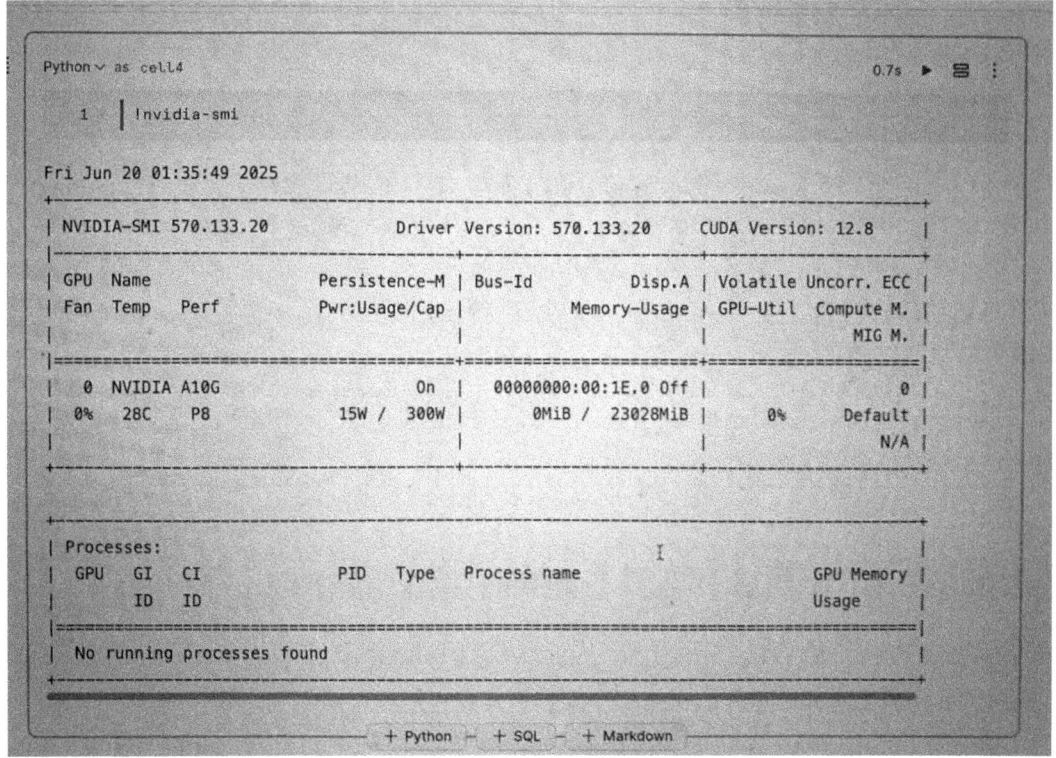

Figure 12-3. *NVIDIA GPU resource usage*

Use the following query to retrieve Snowpark container service credit usage.

```
SELECT
  START_TIME::DATE AS START_DATE,
  APPLICATION_NAME,
  SUM(CREDITS_USED) AS TOTAL_CREDITS_USED
FROM SNOWFLAKE.ACCOUNT_USAGE.SNOWPARK_CONTAINER_SERVICES_HISTORY
WHERE START_TIME > DATEADD(day,-7,CURRENT_TIMESTAMP())
GROUP BY ALL
ORDER BY TOTAL_CREDITS_USED DESC;
```

The following query retrieves the daily credits that are billed for the Snowpark container service.

```sql
SELECT
    USAGE_DATE,
    CREDITS_USED,
    CREDITS_BILLED
FROM SNOWFLAKE.ORGANIZATION_USAGE.METERING_DAILY_HISTORY
WHERE USAGE_DATE >= DATEADD(month,-1,CURRENT_TIMESTAMP())
    AND CREDITS_USED > 0
    AND SERVICE_TYPE = 'SNOWPARK_CONTAINER_SERVICES'
ORDER BY USAGE_DATE;
```

The next section explores one of the most critical areas for optimization, especially when working with LLMs—token consumption,.

Cost Insight: Forecast Snowflake Usage

The *forecast model* is part of Snowflake's native ML functions, allowing users to build, train, and deploy models without moving data outside the platform. The forecast model stands out for making predictive analytics accessible through simple SQL. By analyzing historical data, the model predicts future values and selects the best-fit algorithm automatically. It also supports external variables and provides confidence intervals to show the range of possible outcomes.

In this example, the model is applied to forecast warehouse usage using data from the `ORGANIZATION_USAGE.USAGE_IN_CURRENCY_DAILY` view.

Forecasting warehouse usage is a powerful way to plan budgets, optimize resource allocation, and avoid surprises in your Snowflake billing. In this section, we walk through how to use ML forecasting functions to predict future warehouse usage based on historical spend data.

Understanding the Data

Snowflake tracks daily usage costs in the `USAGE_IN_CURRENCY_DAILY` view. Each row represents the cost incurred by a specific warehouse on a given day.

Here is a quick peek at what the data looks like:

TS	USAGE_COST
2025-01-16	87.40
2025-01-17	119.80
2025-01-18	109.00
2025-01-19	105.60
2025-01-20	105.30
2025-01-21	101.00

We use this data to forecast future daily spend for a warehouse.

Step 1: Prepare the Data

To start, aggregate the daily usage for a specific warehouse. You can modify this to include multiple warehouses or filter it by region/account.

```
CREATE OR REPLACE TEMP TABLE daily_warehouse_usage AS
SELECT
  USAGE_DATE AS TS,
  SUM(USAGE_IN_CURRENCY) AS USAGE_COST
FROM ORGANIZATION_USAGE.USAGE_IN_CURRENCY_DAILY
WHERE WAREHOUSE_NAME = 'WH_USER_M'
AND USAGE_DATE >= DATEADD(day,-180,CURRENT_TIMESTAMP())
GROUP BY ALL;
```

Step 2: Create the Forecast Model

Next, you use the FORECAST model to create a forecast of next 30 days of warehouse usage. All you need is a timestamp column and a target column (in this case, usage_cost).

```
CREATE OR REPLACE SNOWFLAKE.ML.FORECAST warehouse_usage_fc (
  INPUT_DATA => TABLE(SELECT TS, USAGE_COST FROM daily_warehouse_usage),
  TIMESTAMP_COLNAME => 'TS',
  TARGET_COLNAME => 'USAGE_COST'
);

call warehouse_usage_fc!FORECAST(FORECASTING_PERIODS => 180);
```

This will return a table with the forecasted values and confidence intervals:

TS	FORECAST	LOWER_BOUND	UPPER_BOUND
2025-06-16	104.70	93.15	116.26
2025-06-17	108.91	97.32	120.50
2025-06-18	100.66	89.02	112.30
2025-06-19	103.06	91.38	114.74
2025-06-20	110.57	98.85	122.29
2025-06-21	104.47	92.71	116.23

Step 3: Save and Analyze the Forecast

You can now join this with your historical data to compare actual vs. forecasted usage.

With just a few SQL commands, you can build a reliable forecast of your Snowflake warehouse usage. This helps with budgeting, capacity planning, and understanding usage trends, all without needing to be a data scientist. Snowflake ML functions make it easy to get started, and you can always refine your model by adding more features or adjusting the forecast horizon.

```
CREATE OR REPLACE TABLE warehouse_forecast_output AS
    SELECT ts::DATE AS TS
        , forecast
        , lower_bound
        , upper_bound
    FROM (SELECT * FROM
        TABLE(warehouse_usage_fc!FORECAST(FORECASTING_PERIODS => 90)))
    GROUP BY ALL;
```

You can compare the historical and forecasted numbers by plotting them in a graph, as shown in Figure 12-4.

```
WITH actual AS
    (SELECT TS
        , USAGE_COST
        , NULL AS forecast
    FROM daily_warehouse_usage)
    , forecasted AS
```

```
      (SELECT TS
            , NULL AS cnt
            , forecast
        FROM warehouse_forecast_output)
SELECT *
FROM actual
UNION ALL
SELECT *
FROM forecasted
ORDER BY TS;
```

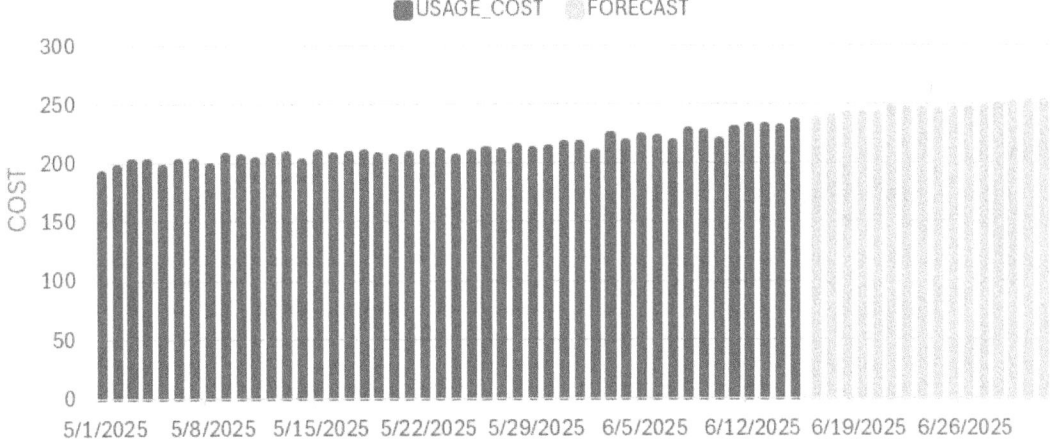

Figure 12-4. Snowflake usage forecast report

Expanding on these features, Snowflake provides innovative tools that simplify advanced analysis and make it accessible to users through Streamlit.

Cost Insight: LLM-Powered Queries

Cortex Analyst is a fully-managed, LLM-powered Snowflake Cortex feature that helps you create applications capable of reliably answering business questions based on your structured data in Snowflake. With Cortex Analyst, business users can ask questions in natural language and receive direct answers, all without writing SQL. Available as a convenient REST API, Cortex Analyst can be seamlessly integrated into any application.

CHAPTER 12   SNOWFLAKE AI/ML COST MANAGEMENT

In this section, we explore how to use Cortex to enable natural language queries for cost-related data. This approach makes cost analysis more accessible to nontechnical users and enhances self-service analytics.

Source code for the Cost Management app is available at https://github.com/velunatsf/sf-llm-cost-governance.git.

Example Use Cases

Using LLMs and tools like Streamlit, users can ask questions such as:

- "What are the total costs in dollars for the organization, broken down by account?"
- "What was my total storage cost for the last month?"

These capabilities empower users to gain insights quickly and intuitively, without needing to understand the underlying data structures or write complex queries.

Find organization level cost usage by account:

	ACCOUNT_NAME	USAGE_DATE	USAGE_IN_CURRENCY
0	YH91765	2024-12-14	1.67
1	YH91765	2024-12-15	3.58
2	YH91765	2024-12-16	0.01
3	YH91765	2024-12-17	12.39
4	YH91765	2024-12-18	3.61
5	YH91765	2024-12-19	20.48
6	YH91765	2024-12-20	15.41
7	YH91765	2024-12-21	23.43
8	YH91765	2024-12-22	4.97
9	YH91765	2024-12-23	0.1

A — This is our interpretation of your question: What are the total costs in dollars for the organization, broken down by account?

CHAPTER 12 SNOWFLAKE AI/ML COST MANAGEMENT

Find storage usage by date:

CHAPTER 12　SNOWFLAKE AI/ML COST MANAGEMENT

This overview of Cortex Analyst and its use in cost management highlights how these capabilities align with the broader set of AI and ML tools available in Snowflake. Together, they support better cost control and more efficient resource planning.

Summary

AI and ML capabilities available in Snowflake provide powerful tools for natural language processing, document analysis, and intelligent automation, all within the data platform. This chapter explained how these features are priced, focusing on token usage and compute resources, and it offered strategies for managing costs effectively. It also introduced tools for monitoring usage and forecasting spend to help teams align AI investments with business goals.

The next chapter highlights Snowflake's newest advancements in cost control, including faster Gen2 warehouses, intelligent Adaptive Compute, reduced replication egress costs, and streamlined operations with the ability to cancel all queries. These are all designed to improve performance, simplify management, and optimize spend across workloads.

CHAPTER 13

What's New in Snowflake Cost Optimization

Overview

This chapter explores the latest advancements in cost optimization from Snowflake, showcasing how the platform continues to evolve to deliver greater value for every dollar spent. The focus is on architectural innovations and intelligent automation that simplify cost governance while enhancing performance. Moving beyond manual tuning and static configurations, Snowflake introduces dynamic, AI-enhanced capabilities that optimize resource use in real time. These developments reflect a strategic shift toward proactive, scalable, and transparent cost management, empowering organizations to align their data operations with financial objectives without compromising agility or performance.

The latest advancements in cost optimization covered in this chapter include:

- How the AI-driven architecture in Snowflake enables real-time cost optimization across diverse workloads.

- The performance and cost-efficiency benefits introduced by next-generation and adaptive compute models.

- Methods for using automated anomaly detection to help FinOps teams proactively manage unexpected spend.

- Strategies to reduce data replication costs in multi-cloud environments.

- Innovations that support scalable, cost-aware data operations across the enterprise.

To begin exploring these advancements in depth, the following section examines how intelligent automation is reshaping cost efficiency across the Snowflake platform

Cost Efficiency Through Snowflake Cortex AI

Snowflake Cortex AI introduces a transformative approach to cost management by embedding intelligence directly into the data platform architecture. This integration enables organizations to optimize resource usage, streamline operations, and make smarter decisions, all while maintaining cost control.

By leveraging AI to drive smarter, more efficient decision-making across the data lifecycle, Cortex AI reduces infrastructure and operational overhead and enhances overall productivity. As AI capabilities continue to evolve, Snowflake positions Cortex AI at the forefront of the data cloud revolution, where cost management is no longer just about cutting expenses, but about maximizing value through intelligent, automated insights at every layer of the architecture.

This streamlining of data operations translates directly into measurable cost savings. Moreover, Snowflake architecture is designed to be AI-ready, supporting both structured and unstructured data within a unified governance framework. This eliminates the need for separate systems or data silos, significantly reducing infrastructure complexity and associated costs.

Figure 13-1 illustrates the Snowflake AI data cloud platform, highlighting how Cortex AI is embedded across the architecture to enable intelligent, cost-efficient data operations.

CHAPTER 13 WHAT'S NEW IN SNOWFLAKE COST OPTIMIZATION

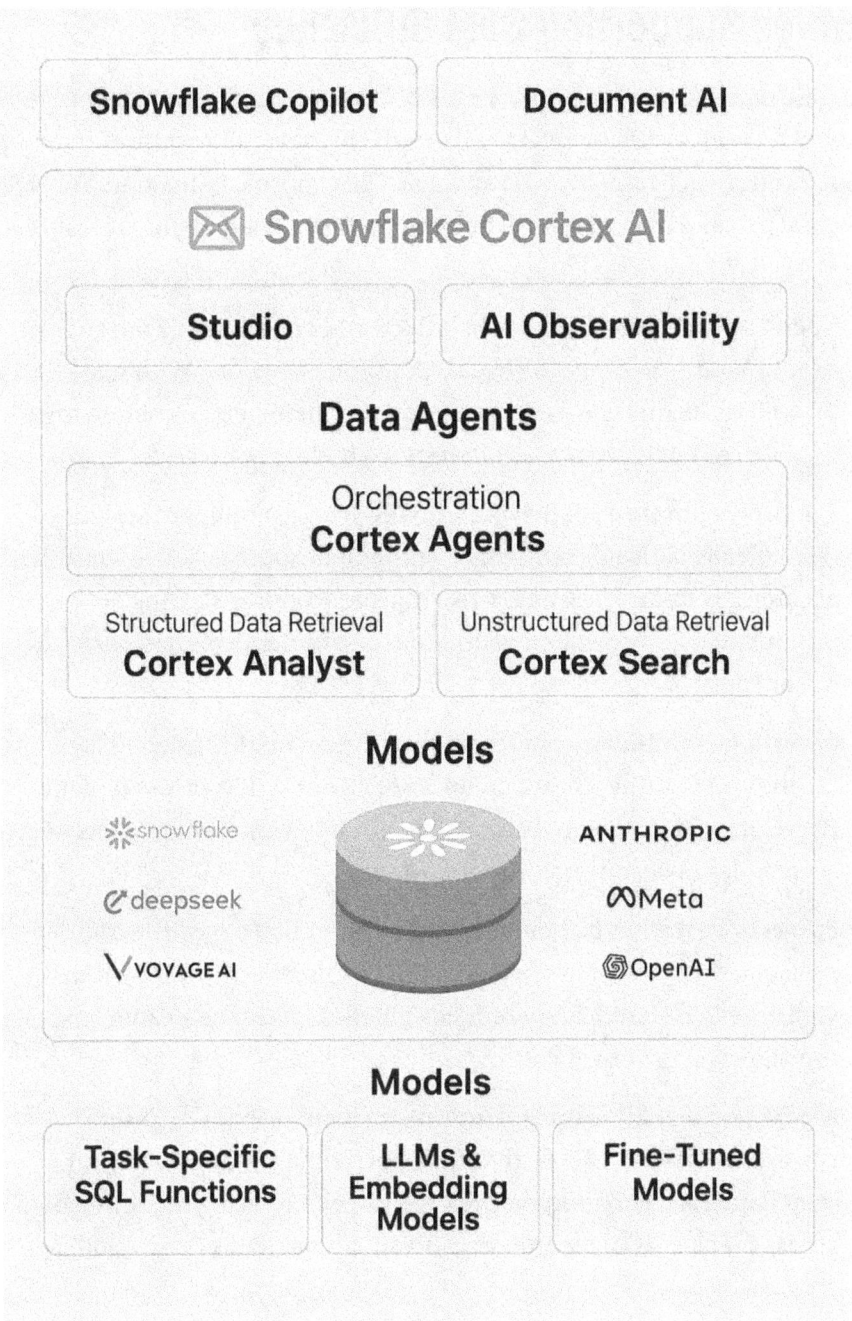

Figure 13-1. Snowflake AI data cloud platform (image courtesy Snowflake Inc.)

AI Features Supporting Cost Efficiency

One of the most transformative applications of Cortex AI in the Snowflake ecosystem is its ability to drive cost efficiency through a suite of intelligent features that optimize performance and reduce operational expenses. This approach illustrates how Snowflake is evolving, not just as a data platform, but as a partner in helping organizations work smarter.

- **Spend anomaly detection (GA):** Cortex AI continuously monitors usage patterns to detect unusual spending behaviors. With real-time alerts, teams can proactively address inefficiencies before they escalate, reducing waste and improving budget adherence.

- **Adoptive compute (in private preview):** This intelligent service dynamically allocates right-sized compute resources across an account. By optimizing query routing and resource sharing, it minimizes over-provisioning and maximizes performance per dollar spent.

- **Snowflake intelligence (soon in public preview):** Designed for business users, this feature enables secure, no-code access to data. By reducing reliance on technical teams, it lowers support costs and accelerates decision-making.

- **Snowconvert AI:** A free, automated solution that simplifies and accelerates data warehouse, BI, and ETL migrations, as shown in Figure 13-2. It significantly reduces the time, complexity, and cost of transitioning to Snowflake.

- **AISQL performance optimizations (private preview):** Enterprises can experience 30–70% faster query performance, depending on their datasets. Even more impressively, when filtering or joining data across thousands of records, organizations can achieve up to 60% cost savings.

- **AI observability (GA soon):** A suite of tools for evaluating, debugging, and optimizing generative AI applications. These capabilities help teams maintain performance and avoid costly inefficiencies in AI workloads.

Snowconvert is available as a standalone tool for download.

Figure 13-2. Snowconvert AI

All these capabilities are built to continuously optimize resource utilization and minimize operational inefficiencies, enabling organizations to better control spending and realize measurable cost savings over time.

Note Feature availability is based on the Snowflake release state as of June, 2025.

Expanding on this foundation of AI-driven cost efficiency, the next innovation introduces advancements in compute performance through the evolution of Gen2 standard warehouses.

Gen2 Standard Warehouses

The release of generation 2 standard warehouses (Gen2) represents a significant evolution in the virtual warehouse architecture aimed at delivering faster query performance, greater concurrency, and improved cost optimization for analytics and data engineering workloads.

What Are Gen2 Standard Warehouses?

Gen2 standard warehouses are the next generation of standard virtual warehouses in Snowflake. Built on faster underlying hardware and enhanced with intelligent software optimizations, they improve the efficiency of common operations such as:

- DELETE, UPDATE, and MERGE operations
- Table scans
- Concurrent query execution

These enhancements allow users to do more work in less time, which directly translates to lower compute costs and better resource utilization.

Key Benefits for Cost Governance

From a FinOps perspective, Gen2 warehouses offer several compelling advantages:

- **Improved performance-to-cost ratio:** With Gen2, most queries complete faster, meaning that fewer credits are consumed per workload. This is especially beneficial for organizations with high query volumes or complex transformations.

- **Higher concurrency:** Gen2 supports more concurrent queries without requiring larger warehouse sizes. This reduces the need to scale up warehouses just to handle peak loads, helping teams avoid overprovisioning.

- **Optimized resource allocation:** The new RESOURCE_CONSTRAINT clause allows users to explicitly choose between STANDARD_GEN_1 and STANDARD_GEN_2 when creating or altering a warehouse. This gives teams more control over performance tuning and cost management.

- **Seamless migration:** Existing standard warehouses can be converted to Gen2 by suspending the warehouse and updating the RESOURCE_CONSTRAINT setting. This flexibility allows teams to test and adopt Gen2 gradually, without disrupting production workloads.

Gen2 Warehouse Credit Table

While Gen2 offers significant performance advantages, organizations should carefully weigh these benefits against the higher operational costs to ensure an optimal balance between performance and cost efficiency. As shown in Table 13-1, on average, Gen2 warehouses consume approximately 1.35 times more credits on AWS and GCP, and 1.25 times more on Azure compared to Gen1.

Table 13-1. Gen2 Warehouse Credit Table (Source: Snowflake Inc.)

Snowflake Warehouse Size	Gen1 Credits/Hour	Gen2 Credits/Hour (AWS and GCP ~1.35x)	Gen2 Credits/Hour (Azure ~1.25x)
X-Small	1	1.35	1.25
Small	2	2.70	2.50
Medium	4	5.40	5.00
Large	8	10.80	10.00
X-Large	16	21.60	20.00
2X-Large	32	43.20	40.00
3X-Large	64	86.40	80.00
4X-Large	128	172.80	160.00
5X-Large	256	N/A	N/A
6X-Large	512	N/A	N/A

Source: Snowflake Inc. Credit consumption table: https://www.snowflake.com/legal-files/CreditConsumptionTable.pdf

For FinOps practitioners, Gen2 warehouses introduce a powerful lever to enhance efficiency, control costs, and maximize the return on investment (ROI) of data workloads. To make informed decisions, it is strongly recommended to benchmark Gen2 performance against existing workloads, allowing teams to quantify the benefits within their specific operational context.

Figure 13-3 illustrates the provisioning process for Gen2 warehouses, emphasizing how configuration flexibility enables gradual adoption.

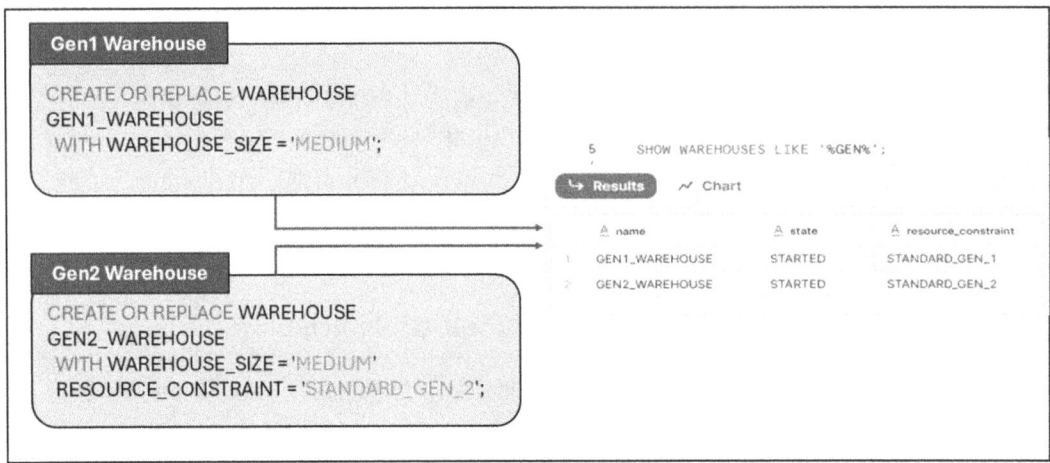

Figure 13-3. Snowflake gen-2 warehouse

Note At the time of writing, the STANDARD_GEN_2 setting is not yet available in the Snowsight UI. Configuration must be performed via SQL using the CREATE/ALTER WAREHOUSE commands.

While Gen2 standard warehouses enhance performance through architectural improvements, the next evolution focuses on intelligent workload management, introducing a dynamic compute model that adapts in real time to query demands.

Adaptive Compute for Smarter Workload Management

At the 2025 Snowflake Summit, Snowflake introduced a transformative innovation in its compute architecture—*adaptive compute,* also known as *adaptive warehouses.* This new compute model is designed to intelligently route queries to the most appropriate warehouse size, eliminating the need for manual sizing and significantly improving performance and cost efficiency.

This advancement represents a major shift in how organizations interact with the compute layer, moving from static, manually configured warehouses to a dynamic, plan-aware system that adapts in real time to the specific needs of each query.

What Is Adaptive Compute?

Traditionally, Snowflake users had to select a fixed warehouse size (e.g., X-Small to 6X-Large) and configure scaling policies to handle concurrency and performance. This required deep knowledge of workload patterns and often led to overprovisioning or underperformance.

Adaptive compute changes this paradigm. As illustrated in Figure 13-4, instead of selecting a warehouse size, users simply submit queries, and Snowflake's intelligent routing engine automatically assigns the query to a warehouse that best matches its resource needs. This is based on the query plan, including operator mix, row estimates, and expected execution time as stated. This provides the following benefits:

- **Plan-aware query routing:** Queries are matched to the optimal warehouse size based on their complexity.

- **Dynamic resource allocation:** No need to predefine warehouse sizes or scaling policies.

- **Simplified configuration:** Users only need to define maximum credits per hour and maximum warehouse size per statement.

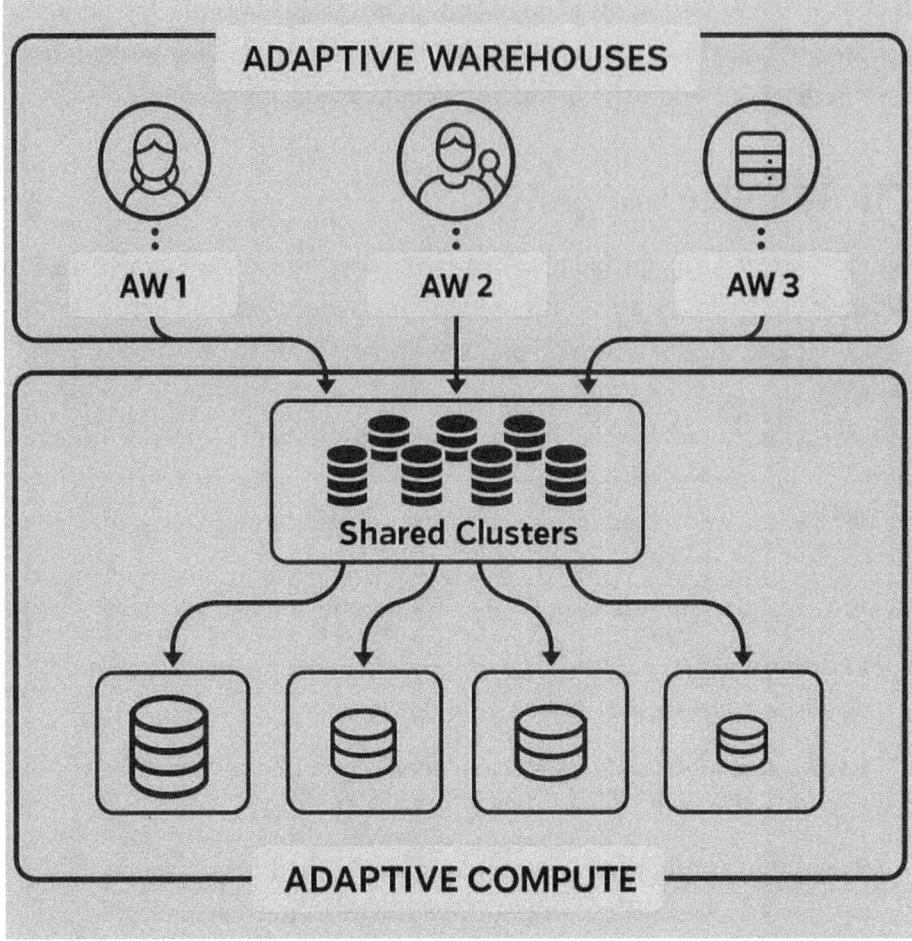

Figure 13-4. Adaptive compute architecture (image courtesy Snowflake Inc.)

Why Adaptive Compute Matters for FinOps

From a FinOps perspective, adaptive compute introduces a new level of automation and efficiency in cost management, as illustrated in Figure 13-5. Here is how it improves the following:

- **Operational simplicity:** Adaptive compute removes the need for manual warehouse sizing, auto-suspend tuning, and multi-cluster configuration. This reduces engineering overhead and allows teams to focus on service-level objectives (SLOs) rather than on infrastructure management.

- **Optimized cost-to-performance:** By routing each query to the most efficient warehouse, Snowflake minimizes credit consumption without sacrificing performance. This is especially valuable for mixed workloads, where some queries are lightweight and others are resource-intensive.

- **Reduced queue times:** A shared pool of warehouses managed by Snowflake ensures that queries are less likely to be delayed due to resource contention. This improves user experience and reduces the need for overprovisioning to handle peak loads.

- **Better budget predictability:** With the ability to cap credits per hour, FinOps teams gain more control over spend while still benefiting from dynamic performance scaling.

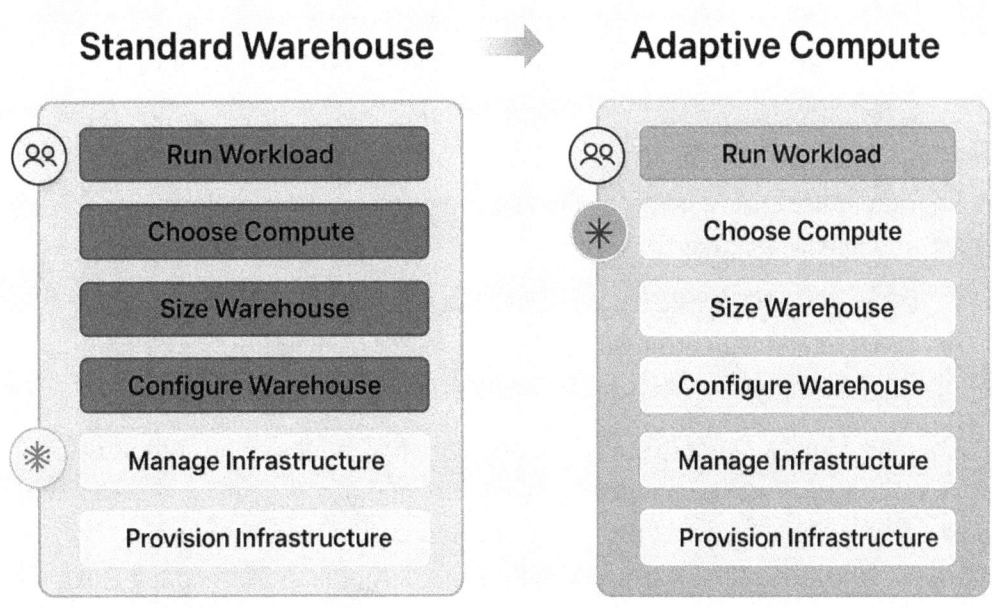

Figure 13-5. Adaptive compute management (image courtesy Snowflake Inc.)

Real-World Impact: The Experience of Pfizer

In the same Summit, Pfizer shared how adaptive compute revolutionized their data operations. By adopting adaptive warehouses, Pfizer was able to:

- Reduce query latency by over 40%
- Eliminate manual warehouse tuning
- Improve cost transparency and forecasting

This shift allowed their data teams to focus on delivering insights rather than on managing infrastructure.

Considerations and Limitations

While adaptive compute offers significant benefits, it is important to understand its current limitations:

- **Private preview:** As of mid-2025, adaptive compute is in private preview and not yet generally available.
- **No manual overrides:** Users cannot force a query to run on a smaller warehouse to save costs. The routing engine prioritizes performance and SLA adherence.
- **Pricing model:** While the system is designed to be cost-efficient, cost savings are not guaranteed in all scenarios. Some workloads may be routed to larger warehouses to meet performance targets.

For organizations looking to scale data operations without scaling complexity, adaptive compute is a game-changer. As the feature matures, we can expect even more intelligent features that blend AI-driven optimization with cost governance.

The next innovation focuses on automated anomaly detection, enabling FinOps teams to identify unexpected spending patterns before they impact budgets or operations.

Spend Anomalies with Real-Time Notification

The new *cost anomalies* feature introduces a powerful, AI-driven capability that automatically detects unusual patterns in Snowflake spend. This innovation helps FinOps teams identify and respond to unexpected cost spikes or drops before they impact budgets or operations.

What Are Cost Anomalies?

Cost anomalies are unexpected deviations in Snowflake credit consumption that differ significantly from historical usage patterns. These could be caused by:

- A misconfigured or runaway query
- A sudden increase in user activity
- A change in data pipeline behavior
- An unplanned workload or integration

Snowflake now uses machine learning algorithms to analyze historical consumption data and automatically flag these anomalies. This allows teams to detect issues early, investigate root causes, and take corrective action, without needing to manually monitor dashboards or set static thresholds. That includes:

- **Automated detection:** Snowflake continuously monitors your account and organization-level usage, comparing current consumption against historical baselines. When a significant deviation is detected, it flags the anomaly for review.

- **Multi-level visibility:** Anomalies can be detected at both the account and organization levels, giving FinOps teams a comprehensive view of where and how costs are changing.

- **Contextual insights:** Each anomaly includes metadata such as:
  - Date and time of occurrence
  - Magnitude of deviation
  - Affected services or warehouses
  - Historical comparison data

This context helps teams quickly assess whether the anomaly is expected (e.g., a planned data load) or requires intervention.

- **Snowsight integration:** The feature is integrated into Snowsight, making it easy to view and investigate anomalies alongside other cost metrics, as shown in Figure 13-6.

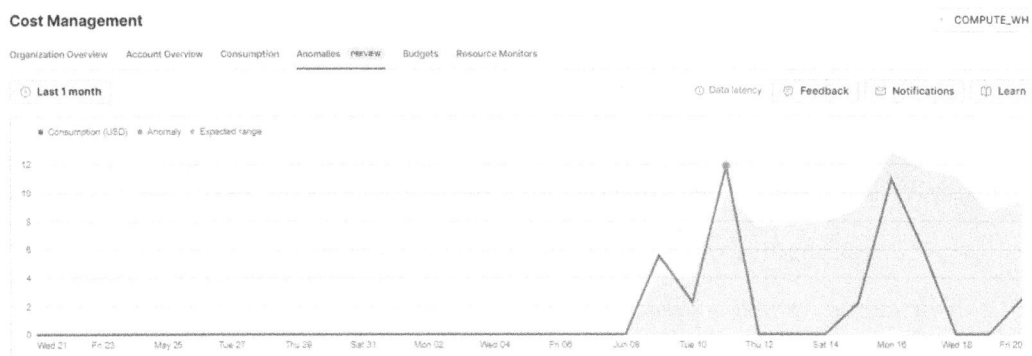

Figure 13-6. Cost management - anomalies

Example Use Case

Imagine a scenario where a data engineering team accidentally deploys a new ETL job that runs every five minutes instead of once per hour. This causes a 10-fold increase in compute usage overnight. Without anomaly detection, this spike might go unnoticed until the monthly bill arrives.

With the cost anomalies feature, the system detects the spike within hours, flags it in Snowsight, and alerts the FinOps team. They investigate, identify the misconfigured job, and correct it, saving thousands of dollars in potential waste.

Currently, this feature is in preview, and Snowflake recommends using it alongside existing monitoring tools for the best results. By automating anomaly detection, Snowflake empowers FinOps teams to shift from reactive to proactive cost management, a critical step in scaling cloud financial operations.

The next section introduces *egress replication cost optimization*, a feature that reduces data transfer costs while maintaining performance and scalability.

Cost Management: Cross Account Cost View

The recently introduced GLOBALORGADMIN role is central to managing and overseeing cost-related activities across an entire organization. This role is designed for users operating within the organization account, providing them with comprehensive visibility and control over usage and billing data across all Snowflake accounts under the organization.

- **Access to organization-wide cost data:** Users with the GLOBALORGADMIN role can view cost and usage metrics across all accounts in the organization. This includes:
  - Total credit consumption
  - Billing summaries
  - Contract balances (when currency view is enabled)
  - Usage trends and forecasts

- **Visibility into currency-based metrics:** When signed in with the GLOBALORGADMIN role, users can view cost data not just in Snowflake credits but also in actual currency values. This is particularly useful for financial planning and contract management.

- **Centralized monitoring and reporting:** The role allows access to the Organization Overview tab in Snowsight, where administrators can monitor:
  - Account-level usage
  - Cost spikes or anomalies
  - Resource consumption patterns

- **Delegating cost management access:** The GLOBALORGADMIN role can grant cost-related access to other users by assigning them appropriate application and database roles, such as:
  - APP_USAGE_VIEWER/USAGE_VIEWER for read-only access
  - APP_USAGE_ADMIN/USAGE_ADMIN for administrative access

- **Security and governance:** By centralizing cost visibility, organizations can enforce consistent governance policies, ensuring that only authorized users can access sensitive financial data.

This role is essential for organizations with multiple Snowflake accounts, as it provides a unified interface for managing resources and policies. The Organization Overview can be accessed through Cost Management, as illustrated in Figure 13-7.

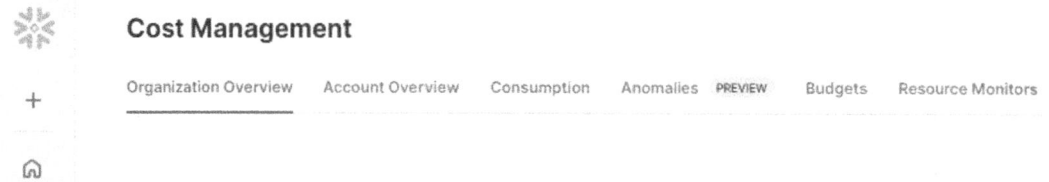

Figure 13-7. Organization cost management

Now that a clear view of organizational costs is in place, the next step is optimizing one of the most significant cost drivers in multi-cloud environments: egress replication using a recently released enhancement.

Egress Replication Cost Optimization

In multi-cloud environments, managing data replication costs has become a critical concern. Snowflake's *Egress Cost Optimizer* (ECO), introduced as part of its cross-cloud auto-fulfillment framework, directly addresses this challenge by minimizing egress charges during cross-region and cross-cloud data sharing. ECO intelligently analyzes data listing configurations and leverages a Snowflake-managed cache to avoid redundant data transfers. This approach delivers substantial cost savings, particularly for data providers operating on a scale.

Seamlessly integrated with Snowflake's security and replication features, ECO ensures cost-efficiency without compromising performance or operational integrity. It represents a strategic advancement in optimizing cloud data economics, enabling scalable, cost-effective data distribution across regions and cloud platforms.

Note For a detailed discussion on ECO and its implementation, refer to Chapter 11.

Summary

Snowflake demonstrates a strong commitment to redefining cost efficiency through intelligent platform enhancements. Recent innovations reduce operational overhead, enhance workload performance, and offer deeper visibility into spending patterns. These capabilities are designed to empower FinOps teams with real-time insights that help balance performance and cost. By embedding optimization into the core architecture, Snowflake enables organizations to scale data operations with confidence and sustainability.

As emphasized throughout this book, cost management is no longer a reactive process. It has become a strategic, built-in capability of the modern data cloud. Cost governance must be embedded in every design and development decision to ensure that business value is maximized for every dollar spent on Snowflake.

Index

A

ACCESS_HISTORY, 158
Adaptive compute
    architecture, 356
    benefits, 355
    FinOps, 356, 357
    limitations, 358
    real world impact, 358
Adaptive warehouses, 354
Artificial intelligence (AI), 319
AWS SNS (Simple Notification Service), 257
Azure Blob Storage, 257

B

Business continuity plan (BCP), 71, 296

C

Capital expenditure (CapEx) model, 117
Chargeback and showback models, 47
Chargeback model, 102
Cloud data platforms, 7
Cloud financial management, 3
Cloud services, 6, 66
Cloud services adjustment, 67
Cloud services optimization
    architecture, 278, 280
    data dictionary (DD) queries
        ACCOUNT_USAGE, 292
        create copy, 293
        periodic data dictionary cleanup, 295, 296
        SHOW commands, 294
        storage costs, 292
    GSL, 277
    metadata layers, 286–288
    metadata operations, 288–291
    monitoring usage, 280–282
    operational efficiency/cost-effectiveness, 277, 278
    patterns, 283–286
    reducing metadata operations, 291
Clustering
    benefits, 216
    cost metrics, 217
    data load operations, 211
    estimate cost, 211–213
    key usage, 214, 216
    query latency, 210
    track table updates, 213
Compute optimization strategies
    attribute cost, 136
    MPP compute, 116, 117
    resource utilization, warehouse, 139
        auto-clustering, 149, 150
        cache, 147, 148
        disk spilling, 143, 144
        high scanners, 145, 146
        OOM, 144, 145
        row explosion, 139–142
        UNION without ALL, 151, 152
    warehouse cost, 137, 138

INDEX

Compute optimization strategies (*cont.*)
   warehouse provisioned
      capacity, 131–133
   warehouse sizing, 118
   warehouse utilization heatmap, 134, 135
Consumption pricing model
   businesses manage resources, 9, 10
   challenges, 10, 11, 13
   pros/cons, 13–16
Content automation, 58
Continuous data protection (CDP), 178
Control phase
   access control
      built-in tools, 112
      granular permissions, 111
      least privilege/regular permission review, 112
      practices, 113
      RBAC, 110
   budget and business goals, 91
   chargeback/snowback models, 102–104
   cloud cost governance
      account budgets, 99
      advanced controls, 101
      budget, 97, 98
      budgeting/alerts, 98
      budgeting management, 100
      budgeting process, 99
      custom budgets, 100
      spending limits/alerts, 100
   monitoring usage/limits
      actions/intelligent alerts, 95
      common pitfalls, 97
      credit quotas, 93
      resource monitors, 93
      scalability and performance, 92
      sustainable monitoring, 96
      types/granularity, 93, 94
   performance/cost, 106–109
   warehouse efficiency, 104–106
COPY INTO command, 251
Cortex AI
   cost considerations, 324–327
   features, 323
   tokens, 327–329
   token usage comparison, 330, 331
Cortex Analyst, 341–344
Cost anomalies
   cost management, 360
   cross account cost view, 361, 362
   ECO, 362
   historical usage patterns, 359
   thresholds, 359
   use case, 360
Cost governance, 363
Cost management
   compute service cost
      AI services, 80
      cloud service, 78
      LLM function, 81
      serverless compute, 79, 80
      virtual warehouse, 77
   data transfer cost, 82
   predefined queries, 77
   priority support, 83
   storage charge, 81
   tagging/cost attribution, 84–87
   UI
      account level, 73, 74, 76
      cost insight, 74, 75
      incurred costs, 76
      organization level usage, 73
      warehouse consumption, 75
   usage statement, 88, 89

Cost optimization
    AI-driven architecture, 347
        cost efficiency, 348–351
Credit quotas, 93

D

Data availability
    ECO, 301–303
    leverage replication/strategic cloud transitions, 315, 316
    replication costs, 300, 301, 316–318
    replication optimization, 304
        audit replication history, 307, 308
        failover expenses, 308, 309
        monitor replication activity, 304, 305
        object selection, 311
        policies, business needs, 309, 310
        set failover testing, 312, 313
        track databse-level replication costs, 305, 306
    RPO and RTO, 298
    service level, 299
    storage costs, secondary account, 313–315
Data Definition Language (DDL) operations, 284
Data lakehouse architecture, 201
Data loading
    batch loading, 254, 255
    Iceberg tables, 263–266
    planning strategies, 249
    snowflake dynamic tables, 259–262
    snowpipe, continuous data ingestion
        benefits, 258
        cloud message integration, 256
        cost model, 259
        micro-batches, 256
        REST API integration method, 257
        SQL example, 258
    strategies, 250–253
    tailoring VWs size
        COPY operations, 267
        COPY_HISTORY and QUERY_HISTORY, 269–271, 273–276
        metadata tables, 267
        methodology, 268
Data modeling optimization
    clustering, 210
    denormalizing fact/dimension tables, 247
    hybrid tables, 232–236
    materialized view, 218, 221–224
    semi-structured data
        dos and don't's, 246
        JSON extraction, 242, 243
        measure performance metrics, 246
        metadata engine, 241, 242
        query performance, JSON data, 243
        usage optimization, 244, 245
    soft referential integrity, 237–241
    style, 209
Disaster recovery (DR), 297
Disk spilling, 143

E

Egress Cost Optimizer (ECO), 301, 362
Extract, Load, Transform (ELT), 5, 168

F

Financial operations (FinOps), 1
    automatic clustering, 51
    benefits, 8, 9
    capabilities, 8

Financial operations (FinOps) (*cont.*)
- control/budgeting phase, 42, 43
  - automation, 45, 46
  - collaboration and communication, 49
  - continuous improvement and feedback loops, 48, 49
  - controls/policies, 44, 45
  - cost accountability/chargeback/showback, 47, 48
  - monitoring/compliance, 46, 47
  - objectives, 44
- framework phases, 18, 19
- materialized views, 52
- operational efficiency, 17
- optimization phase, 50
- phases, 31–33
- QAS, 52
- query caching, 54
- query profiling and optimization, 50, 51
- strategy components, 20–23
- time travel and data retention, 53
- visibility/transparency phase
  - components, 34
  - cost awareness, 38
  - cost monitoring, 36
  - key cost drivers, 35
  - metrics/KPIs, 39
  - organizational culture, 41
  - practices, 40
  - snowflake, 35
  - tagging and governance, 37
- Forecast model, 338

G

Generation 2 standard warehouses (Gen2), 351
- benefits, 352
- credit table, 353, 354
- operations, 352

Global service layer (GSL), 277

H

High availability/disaster recovery (HA/DR) strategies, 298

I, J

Iceberg tables, 263

K

Key performance indicators (KPIs), 20, 39

L

Large language model (LLM), 68, 319
LLM-powered Snowflake Cortex, 341

M, N

Machine learning (ML), 319
Massively parallel processing (MPP), 115
- architecture, 4
- system, 160

Micro-partitions (MPs), 145

O

Object tags, 86
Operational skew
- column, 161
- join/aggregate columns, 162
- mitigate skew, 163
- operations, 160

practices, 161
system-wide impact, 161
Out of memory (OOM), 144

P

"Pay-as-you-go" model, 2

Q

Query Acceleration Service (QAS), 52, 124
Query caching, 54
Query processing, 5
Query tags, 84

R

Recovery point objective (RPO), 298, 299, 301
Recovery time objective (RTO), 298, 299, 301, 312
RELY keyword, 237
REST API integration method, 257
Return on investment (ROI), 57, 207
Role-based access control (RBAC), 110
Row explosion, 139

S

Search Optimization Service (SOS), 213, 224
  benefits, 230, 231
  estimated costs, 225, 226
  guidelines and best practices, 225
  usage metrics, 231, 232
  usage optimization, 227, 228
Service costs
  compute service, 65, 67–69
  data transfer, 71
  primary pricing factors, 64
  priority support, 72
  storage costs, 69, 70
  usage report, 64
Service-level objectives (SLOs), 356
Shared accountability model, 24, 25
SHOW commands, 294
Showback model, 103
Snowflake
  architecture
    centralized data repository, 4
    cloud services, 6, 7
    database storage, 5
    query processing, 5
  architecture and consumption-based pricing model, 1
  costs optimization framework, 26–29
Snowflake AI
  AISQL features, 320
  credit usage, 331–333
  features credit table, 322
  LLMs, 320
  tokens, 322
Snowflake Cortex AI, 348
Snowflake ML
  building cost effective models, 334–338
  forecast model, 338–341
  functions and cost management, 334
Snowflake's zero-copy cloning, 189
Snowflake warehouse, 106, 107
Snowpark, 164, 176
Snowpark Container Services (SPCS), 67
Snowpipe, 255
Soft referential integrity, 237
Storage optimization
  CDP, 178–180

Storage optimization (*cont.*)
   costs, 203, 204
   database usage, 204
   data lakehouse architecture, 201, 202
   design considerations, 177
   large, high-churn tables, 185, 186
   manage database storage costs, 178
   manage storage lifecycle, 197, 198, 200
   point-in-time cloning, 189
   pre-build queries, 205, 206
   reclustering, 187
   short-lived tables, 184, 185
   snowflake stages, 180, 181, 191–193
   table cleanup, 194–197
   table usage metrics, 204
   time travel/fail-safe, 181–184
   workspace governance/cleanup, 200
Storage pricing, 70
Strategic cloud transitions, 315
Streamlit, 342

T
Token, 68
Tri-secret secure (TSS) encryption, 303

U
User-defined functions (UDFs), 165

V
VARIANT datatype, 241
VARIANT type, 248
Virtual warehouses (VWs), 65, 115, 116, 119, 266
Visibility and transparency
   performance/usage data, 61–63
   pricing model, 60, 61
   reporting/continuous evolution, 58–60
   strategies, 57

W, X, Y
Warehouse sizing guidelines
   auto-suspect settings, adjust, 118, 119
   MAX_CONCURRENCY_LEVEL, 123
   multiple clusters settings, 119, 120
      economy policy, 121, 122
      maximized mode, 122
      standard policy, 120, 121
   QAS, 124–128
   STATEMENT_TIMEOUT_IN_SECONDS parameter, 124
   use data scans, 128–130
Workload optimization
   ACCESS_HISTORY, 158–160
   consumption layer, 172–175
   ELT, 168–170
   ETL, 168–170
   high cost query patterns, identify
      query_hash, 154, 155
      query_parameterized_hash, 155, 156
      query patterns, 157
   Snowpark, 164
      optimized compute node, 165
      optimized warehouses, 165, 166
      performance, 167
   strategies, 153
   transformations, 170–172

Z
Zero-copy cloning, 190